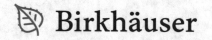

Advances in Mathematical Fluid Mechanics

Lecture Notes in Mathematical Fluid Mechanics

Editor-in-Chief
Giovanni P Galdi
University of Pittsburgh, Pittsburgh, PA, USA

Series Editors
Didier Bresch
Université Savoie-Mont Blanc, Le Bourget du Lac, France

Volker John
Weierstrass Institute, Berlin, Germany

Matthias Hieber
Technische Universität Darmstadt, Darmstadt, Germany

Igor Kukavica
University of Southern California, Los Angles, CA, USA

James Robinson
University of Warwick, Coventry, UK

Yoshihiro Shibata
Waseda University, Tokyo, Japan

Lecture Notes in Mathematical Fluid Mechanics as a subseries of "Advances in Mathematical Fluid Mechanics" is a forum for the publication of high quality monothematic work as well lectures on a new field or presentations of a new angle on the mathematical theory of fluid mechanics, with special regards to the Navier-Stokes equations and other significant viscous and inviscid fluid models.

In particular, mathematical aspects of computational methods and of applications to science and engineering are welcome as an important part of the theory as well as works in related areas of mathematics that have a direct bearing on fluid mechanics.

More information about this subseries at http://www.springer.com/series/15480

Alexander Khapalov

Bio-Mimetic Swimmers in Incompressible Fluids

Modeling, Well-Posedness, and Controllability

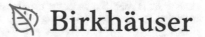 Birkhäuser

Alexander Khapalov
Department of Mathematics and Statistics
Washington State University
Pullman, WA, USA

ISSN 2297-0320   ISSN 2297-0339   (electronic)
Advances in Mathematical Fluid Mechanics
ISSN 2510-1374   ISSN 2510-1382   (electronic)
Lecture Notes in Mathematical Fluid Mechanics
ISBN 978-3-030-85284-9   ISBN 978-3-030-85285-6   (eBook)
https://doi.org/10.1007/978-3-030-85285-6

Mathematics Subject Classification: 35xx, 76xx, 93xx, 49xx

This book is published under the imprint Birkhäuser, www.birkhauser-science.com by the registered company Springer Nature Switzerland AG
The registered company address is: Gewerbestrasse 11, 6330 Cham, Switzerland

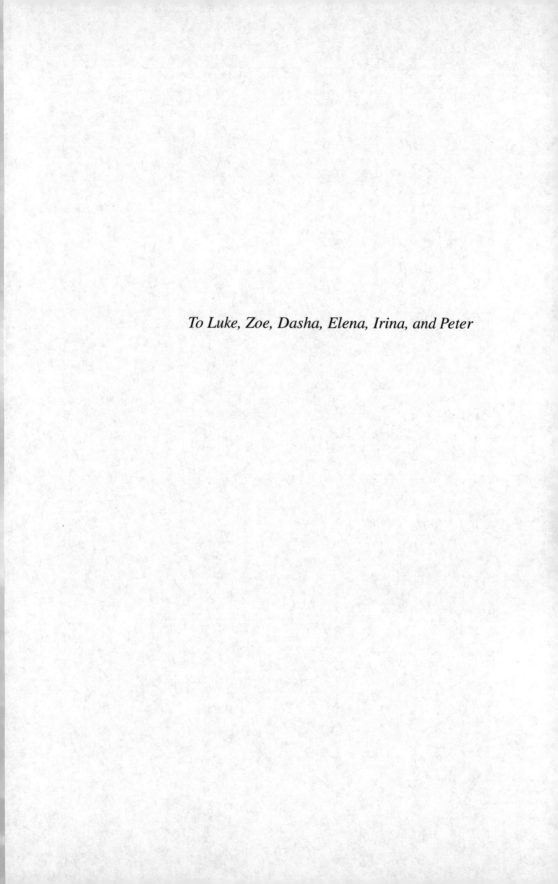

To Luke, Zoe, Dasha, Elena, Irina, and Peter

Preface

The swimming and flying phenomena in nature have been a source of great interest and inspiration for many researchers in mathematics and natural sciences for a long time. The bibliography of formal publications in this area is very diverse in terms of approaches and methodology, and can be traced as far back as to the works of G. Borelli in the seventeenth century.

Along these lines, the goal of this monograph is to present an original concise mathematical theory for so-called "bio-mimetic swimmers" in the framework of coupled system of partial differential equations (PDE) and ordinary differential equations (ODE). This theory includes:

- An original modeling approach (Part I),
- An associated well-posedness results for the proposed models for swimmers (Part II)
- A controllability theory, studying the steering potential of the proposed swimmers (Parts III–V).

We will be focusing on the bio-mimetic swimmers (viewed as possible artificial mechanical devices) created to imitate the swimming locomotion of the existing "swimmers" in nature such as fish, eels, clams, aquatic frogs, turtles, sea snakes, and spermatozoa.

This theory was initially developed in the series of works [3, 19–33], published in 2005–2018, and in this monograph, it is extended to a principally wider class of swimming models with significantly improved steering capabilities.

The proposed theory is based on the original "immerse body" modeling approach, introduced in [19] in 2005. More precisely, we model the interaction between a swimmer and an incompressible fluid surrounding it by making use of a coupled system of two sets of partial and ordinary differential equations:

- A fluid (Navier–Stokes or non-stationary Stokes) equation for the fluid dynamics
- An ordinary differential equation for the position of a swimmer in the fluid

It is assumed that the swimmer's body consists of finitely many elements, identified with the fluid they occupy, that are subsequently linked by the rotational and structural forces, which are explicitly described and serve as the means to change the geometric shape of the swimmer's body. The sum of such forces, being internal to the swimmer at hand, is equal to zero, and hence, they cannot move its center of mass in a space containing no medium. However, in an incompressible fluid, the original swimmer's internal forces, in general, will change, both in their magnitude and direction, according to the instantaneous spatial orientation of the swimmer's body parts and of these forces. Respectively, the sum of the actual forces that will act on the swimmer's body parts in an incompressible fluid can become non-zero, which will result in the swimmer's locomotion (a self-propelling motion).

Acknowledgments

The author's research presented in this monograph was supported in part by the NSF Grants DMS-0504093 and DMS-10007981 and Simmons Foundation award number 317297.

The author wishes to express his gratitude to the Department of Mathematics at the University of Rome II "Tor Vergata" and to the Instituto Nazionale di Alta Matematica (Italy) for hospitality throughout his sabbatical in the Spring semester of 2019 during which parts of this monograph were written.

The author also wishes to thank the reviewers for their comments and suggestions.

Pullman, WA, USA   Alexander Khapalov
Fall 2020–Winter 2021

Contents

Chapter 1
Introduction

The swimming and flying phenomena in nature have been a source of great interest and inspiration for many researchers in mathematics and natural sciences for centuries. The bibliography of formal publications in this area is very diverse in terms of approaches and methodology and can be traced as far back as to the works of G. Borelli in the seventeenth century.

Along these lines, the goal of this monograph is to present an original concise mathematical theory for the so-called "bio-mimetic swimmers" in the framework of coupled system of partial differential equations (PDEs) and ordinary differential equations (ODEs). This theory includes

- an original modeling approach (Part I),
- the associated well-posedness results for the proposed models for swimmers (Part II), and
- a controllability theory, studying the steering potential of the proposed swimmers (Parts III–V).

We will be focusing on the bio-mimetic swimmers (viewed as possible artificial mechanical devices) created to imitate the swimming locomotions of the existing "swimmers" in nature such as fish, eels, clams, aquatic frogs, turtles, sea snakes, spermatozoa, etc.

This theory was initially developed in the series of works [3, 19–33], published in 2005–2018, and in this monograph, it is extended to a principally wider class of swimming models with significantly improved steering capabilities.

The proposed theory is based on the original "immerse body" modeling approach, introduced in [19] in 2005. More precisely, we model the interaction between a swimmer and an incompressible fluid surrounding it by making use of a coupled system of two sets of partial and ordinary differential equations:

- a fluid (Navier–Stokes or non-stationary Stokes) equation for the fluid dynamics and
- an ordinary differential equation for the position of a swimmer in the fluid.

© The Author(s), under exclusive license to Springer Nature Switzerland AG 2021
A. Khapalov, *Bio-Mimetic Swimmers in Incompressible Fluids*, Lecture Notes
in Mathematical Fluid Mechanics, https://doi.org/10.1007/978-3-030-85285-6_1

It is assumed that the swimmer's body consists of finitely many elements, identified with the fluid they occupy, that are subsequently linked by the rotational and structural forces, which are explicitly described and serve as the means to change the geometric shape of the swimmer's body. The sum of such forces, being internal to the swimmer at hand, is equal to zero and, hence, they cannot move its center of mass in a space containing no medium. However, in an incompressible fluid, the original swimmer's internal forces, in general, will change, both in their magnitude and in their direction, according to the instantaneous spatial orientation of the swimmer's body parts and of these forces (we refer for more details to Sect. 1.3 below and to Part IV). The sum of the actual forces that will act on the swimmer's body parts in an incompressible fluid can become non-zero, which will result in the swimmer's locomotion (a self-propelling motion).

In the remainder of this introductory chapter, we will outline the main ideas and research strategy employed below in this monograph, making use of an illustrative example of a "bio-mimetic fish."

1.1 Modeling: Mimicking the Nature

In this section, we will discuss some principal aspects of a swimming motion of a fish, see Figs. 1.1, 1.2, 1.3, 1.4 and 1.5.

Fig. 1.1 Swimming motion of fish 1

Fig. 1.2 Swimming motions of fish 2

Fig. 1.3 Swimming motions of fish 3

Fig. 1.4 Swimming motions of a group of fish

A generic fish makes use of change of geometry of its whole body to propel itself in a fluid:

- For "large scale and fast" motions, it bends its body in an S-shaped wave-like pattern (a self-propelling pattern for many other organisms, e.g., for sea snakes) as shown in Figs. 1.1, 1.2 and 1.3, with the tail fin playing an instrumental role in this motion.
- In turn, to stabilize its body in space, or to choose a new direction for forward motion, or for "small scale" motions, etc., a fish can engage all its fins.

Fig. 1.5 A 3D bio-mimetic "fish-like" swimmer: 2 shapes

Our modeling strategy in Part I below attempts

1. to analyze the critical patterns of swimming locomotion of a particular biological organism and then
2. to distinguish some selected parts of its body that appear to be instrumental for this locomotion,

with the goal to design, out of similar simplified parts, a bio-mimetic swimmer that can imitate the aforementioned swimming locomotion.

An example of this modeling strategy can be a bio-mimetic swimmer shown in Figs. 1.5–1.6, derived from Figs. 1.1, 1.2, 1.3 and 1.4.

A further rigorous mathematical verification of the proposed bio-mimetic models, namely, their mathematical well-posedness and steering capabilities form the main content of Parts II–V of this monograph (see also Sects. 1.2–1.4 below).

Swimmers Controls   The models proposed in Part I of this monograph include two types of available *explicit* controls for a swimmer:

* *internal control forces* and
* *geometric controls*—a set of available geometric transformations for swimmers' body parts.

The purpose of the aforementioned controls is to ensure that the respective bio-mimetic devices are actually capable of a *self-propelling swimming locomotion* (which is *not* a drifting motion or a motion resulted from being pulled by a "rope," etc.). For further motivating and technical details, we refer to Part I of this monograph.

In our models below, the internal control forces typically act upon the centers of mass of individual swimmer's body parts, while the geometric controls ensure that the respective body parts can rotate about their centers of mass to change their spatial orientation. This aspect of modeling is illustrated in Figs. 1.5–1.6 (with more examples in Part I).

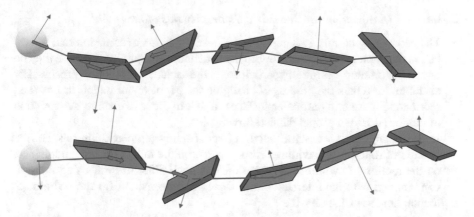

Fig. 1.6 A 3D bio-mimetic "fish-like" swimmer: 2 shapes with some internal structural (or elastic) and rotational forces engaged

In this monograph, for construction of proposed bio-mimetic swimmers, we focus on simple geometric shapes to be used as parts of their bodies, such as parallelepipeds and balls in 3D and discs and rectangles in 2D.

This way, we attempt to simplify the modeling process of our artificial swimmers, while preserving their principal swimming capabilities, as well as to simplify a respective rigorous mathematical analysis of proposed models.

The sum of swimmer's internal forces has to be equal to zero, while their moment has to be conserved, see Fig. 1.6 for illustration. We will elaborate on this aspect in Part I of this monograph below.

Why Internal Forces and Geometric Controls?
The motivation for the introduction of two types of control comes from our modeling strategy to imitate the swimming or flying behavior of living creatures:

- *Examples of geometric controls in nature:*

  1. A bird changes geometry of its wings and tail, by spreading or folding them, or changing their spatial orientation, depending on a desirable flying pattern.
  2. Similarly, a fish can spread or fold a fin or change its 3D spatial form.
  3. Typical swimmers in nature such as frogs, ducks, etc., also change the geometry of some parts of their bodies (say, opening and closing a hand or a web) generating the critical swimming motions with the purpose to either minimize the resistance of the fluid or maximize the area of fluid/body interaction.

- *The internal controls in nature* are usually of rotational type and structural, or elastic, forces such as the forces that make a fish's body to bend, or move a bird's wings, etc.

Remark 1.1 (Assumptions of Geometric Controls in this Monograph)

- The geometric controls, in nature, are the result of action of some internal forces, but, in this monograph, we do not consider such forces explicitly, because they do not, in our opinion, essentially participate in the "actual" locomotion process. For example, for a fish, the process of changing the geometry of its tail or fins (i.e., reorienting them into particular surfaces in 3D, in order to change the direction of motion) does not propel the fish forward.
- In this monograph, we assume that the effect of acting geometric controls, i.e., the process of rotations of swimmer's body parts, on the fluid velocity (and, hence, on the motion of swimmers in this fluid), is negligible. In turn, using them in combination with the internal control forces is instrumental for the swimmers' locomotion, see Chaps. 9–10.

1.2   Mathematical Approach to Swimming Modeling

In this monograph, we follow the "immerse body" modeling approach of [3, 19–33], introduced in [19] in 2005. Namely, we describe the swimmer's locomotion within a given bounded spatial domain Ω in an incompressible 2D or 3D fluid by the following coupled hybrid system of equations:

$$
\begin{cases}
u_t - v\Delta u + (u \cdot \nabla)u + \frac{1}{\rho}\nabla p = \frac{1}{\rho}F(z) & \text{in } Q_T = (0, T) \times \Omega, \\
\operatorname{div} u = 0 & \text{in } Q_T, \\
u = 0 & \text{in } \Sigma_T = (0, T) \times \partial\Omega, \\
u(0, \cdot) = u_0 & \text{in } \Omega,
\end{cases}
\tag{1.2.1}
$$

$$
z(t) = (z_1(t), \dots, z_n(t)),
$$

$$
\frac{dz_i}{dt} = \frac{1}{\operatorname{meas}(S(z_i(t)))} \int_{S(z_i(t))} u(t, x)\, dx, \quad z_i(0) = z_{i,0}, \quad i = 1, \dots, n, \qquad t \in (0, T).
\tag{1.2.2}
$$

In short (for all details we refer to Parts I and II below),

- system (1.2.1) describes the evolution of an incompressible fluid governed by the Navier–Stokes equations (we will also consider the non-stationary Stokes equations), for a given initial fluid velocity u_0, under the influence of the forcing term $F(z_1, \dots, z_n)$ that represents the actions of a given swimmer in Ω on the surrounding medium.

We assume that $x = (x_1, \dots, x_K)$, $K = 2$ or 3, $\partial\Omega$ is a boundary of $\Omega \subset R^K$, $u(t, x)$ and $p(t, x)$ are, respectively, the velocity of the fluid and its pressure at point x at time t, ν is the kinematic viscosity constant, ρ is the fluid density, and we typically assume below, for simplicity, that $\rho = 1$ (see Assumption 2.1 below).

The term $F(z)$ depends explicitly on the swimmer's internal forces and geometric controls applied to respective swimmer's body parts $S(z_i(t))$, $i = 1, \ldots, n$, with centers of mass $z_i(t)$, $i = 1, \ldots, n$, at time t (see Parts I and II and Remark 2.3 below for further details).

- In turn, system (1.2.2) describes the motion of the swimmer within Ω through the motion of the centers of mass $z_i(t)$'s, which allows one to calculate the motion of the swimmer's center of mass.
- The swimmer's flexible body consists of n connected "small" sets $S(z_i(t))$ that can change their spatial orientation in time.
- These sets are identified with the fluid within the space they occupy at time t and are linked between themselves by the rotational and elastic forces acting upon the centers of mass $z_i(t)$'s of the respective separate parts of the swimmer's body (as, e.g., illustrated in Fig. 1.6). The instantaneous velocity of each part is calculated as the average fluid velocity within it at time t.

In Part I below, we will describe several models for various bio-mimetic devices mimicking the locomotion of a fish, a clam, a frog, and similar biological organisms, with numerous illustrations.

In Part II, we will discuss the well-posedness of such models in various function spaces. Namely, we want to know whether a proposed system of equations admits a unique solution in a given function space. We will also discuss continuous dependance of respective solutions with respect to various parameters of a system at hand (the initial fluid, internal swimmer's forces, etc.)

Remark 1.2 (Comment on Modeling Approach) In this monograph we assume that the bodies of swimmers are occupied with the fluid. The motivation for that is due to an empiric understanding that, if the size of a swimmer is negligible relative to the size of swimming domain, this assumption allows one to approximate the actual "real-life swimmers " sufficiently well in terms of the study of their motion from one point to another. For further details and explanations, we refer to the discussion of the *immersed body SIF modeling approach* in Sect. 1.4.

1.3   Swimming Controllability

In Parts III–V of this monograph, we will investigate the steering properties of bio-mimetic swimmers proposed in Part I. Namely, we will investigate whether a particular swimmer at hand is capable of controlled locomotion, relative to its center of mass, from point A to point B within a given fluid domain, while preserving the structural integrity of its body.

We are interested both in local and in global steering properties, as well as in the ability of a swimmer at hand to produce internal force acting upon its center of mass pointed at any desirable direction at any moment of time. Let us elaborate on these aspects of steering.

Swimming Force Controllability When a swimmer engages its internal forces in a fluid, its body parts will press upon (or interact with) the surrounding medium, attempting to overcome its resistance, according to their given geometric shape and instantaneous spatial orientation (i.e., according to geometric controls engaged). Due to the incompressible nature of the fluid, the original swimmer's internal forces, in general, will change, both in their magnitude and in their direction when they act inside of such medium (see Part IV below for concrete technical calculations). In particular, this phenomenon would require an asymmetric shape of swimmer's body parts (see Part IV below for more details). Thus, the sum of such transformed forces can become a non-zero force, which can propel the swimmer at hand in a fluid, or a swimmer's locomotion will occur. (No locomotion occurs under swimmer's internal forces outside any "resisting" medium.)

In Part IV we will investigate the above fluid–swimmer interaction mechanism, namely, how a force acting on a specific shape (in particular, a parallelepiped or ball in 3D or a rectangle or disc in 2D) is transformed into a different force in an incompressible fluid.

Local Controllability In terms of controlled swimmer's locomotion, the question of our interest in Part III is

- *Can a given swimmer move its center of mass from its initial position to any point within some neighborhood of this initial position* (i.e., within a ball in $3D$ or a disc in $2D$)*?*

Global Controllability Here, we would like to know

- *Can a given swimmer move its center of mass from an arbitrary initial position to an arbitrary desirable position within a given fluid domain?*

We approach the answer to this question in Part V below via the following question:

- *Does a given swimmer have an ability,* by engaging its available controls, *to create a force in an incompressible fluid (see Part IV) that can ensure its* momentary *self-propulsion in any desirable direction?*

1.4   Related Selected Bibliography

In this section we will give a brief account of some publications in the area of research for swimming phenomenon in mathematics and natural sciences, as it seems relevant to the subject of this monograph (for a more extended version of such survey, we refer to Khapalov [25], Chapter 10 and also see its short version in [3], and a survey in [11]).

Historically, the first efforts here were aimed at understanding of biomechanics of swimming locomotion of specific species, with principal early contributions due to Gray [12], Gray and Hancock [13], Taylor [54], [55], Wu [58], Lighthill [40], and others.

This research helped to obtain a number of mathematical models for swimming motion, in the *entire R^2-* or R^3-spaces (regarded as a swimming domain), while *the swimmer was regarded as the* reference frame, see, e.g., Childress [4] and the references therein. (The latter means that swimmers' positions in the swimming domain were not explicit.)

Furthermore, based on the size of Reynolds number, it was proposed to divide swimming models into three groups: microswimmers (such as flagella, spermatozoa, etc.) with "insignificant" inertia, "regular" swimmers (fish, dolphins, humans, etc.) whose motion takes into account both viscosity of fluid and inertia, and Euler's swimmers, in which case viscosity is to be "ignored."

It now appears that the following two (in fact, mutually excluding) approaches to model the swimming phenomenon have been in the focus of research interests:

The Shape Transformation Modeling Approach This approach exploits the idea that the swimmer's shape transformations during the actual swimming process can be viewed as a set-valued map in time (see the seminal paper by Shapere and Wilczeck [49]). The respective models describe the swimmer's position in a fluid via the aforementioned maps, see, e.g., [6, 15, 48], and the references therein. Typically, such models consider these maps as *a priori prescribed*, in which case the question whether the respective maps are admissible, i.e., compatible with the principle of self-propulsion of swimming locomotion or not, remains unanswered. In other words, one cannot guarantee that the model at hand describes the respective motion as a self-propulsion, i.e., a swimming locomotion process. To ensure the positive answer to this question, one needs to be able to answer the question whether the a priori prescribed body changes of swimmer's shape can indeed be a result of actions of its internal forces under unknown-in-advance interaction with the surrounding medium.

The Swimmer's Internal Forces Modeling Approach (SIF Approach) In this approach, it is assumed that the available internal swimmer's forces are explicitly described in the model equations and, therefore, they will determine the resulting swimmer's locomotion. In particular, these forces will define the respective swimmer's shape transformations in time as a result of an unknown-in-advance interaction of swimmer's body with the surrounding medium under the action of the aforementioned forces. For this approach, we refer to early principal works by Peskin [46], Fauci and Peskin [9], Fauci [8], Peskin and McQueen [47], Khapalov [25], and the references therein.

The Immersed Boundary SIF Modeling Approach The original idea of Peskin's approach was to view a "narrow" swimmer as an immaterial "immersed boundary." It was proposed to couple a fluid equation with an infinite-dimensional differential equation for the aforementioned immersed boundary.

The following coupled system of Eqs. (1.4.1)–(1.4.3) and figure illustrate this modeling approach (for the case of the fluid density $\rho = 1$), see, e.g., [8]:

$$\frac{\partial u}{\partial t} + u \cdot \nabla u + \nabla p = \nu \Delta u + F(t, x), \quad \nabla \cdot u = 0, \tag{1.4.1}$$

$$F(t, x) = \int f(t, s)\delta[x - X(t, s)]ds. \tag{1.4.2}$$

In (1.4.1)–(1.4.3), the state of the *massless* organism is given by the spatial configuration of points $X(t, s)$, where s is an arc-length parameter; $f(t, s)$ is the boundary force per unit length at each point of the immersed organism, determined by its shape at time t. The integration is over the immersed organism, and δ denotes the two-dimensional delta function. The velocity of each point of the organism is equal to the fluid velocity evaluated at that point:

$$\frac{\partial X(t, s)}{\partial t} = \int u(t, s)\delta[x - X(t, s)]dx, \tag{1.4.3}$$

where the integration is over the entire fluid domain (Fig. 1.7).

The Immersed Body SIF Modeling Approach Motivated by the ideas of Peskin's approach, we introduced the *immersed body SIF modeling approach* in which the bodies of "small" flexible swimmers are assumed to be identified with the fluid within their shapes, see [3, 19–33] (2005–2018). Indeed, say, in 2D, in the framework of Peskin's modeling approach, the swimmer is modeled as an immaterial curve, identified with the fluid, and further discretized for computational purposes on some grid as a collection of finitely many "cells," which in turn can, in fact, be viewed as an immerse body, namely, within the employed computational framework.

The idea of our immersed body modeling approach, employed in this monograph, is to try, making use of mathematical simplifications of such approach, to focus on the issue of macro dynamics of a swimmer. The simplifications (they seem to us to be legitimate within the framework of our interest) include

Fig. 1.7 "Immersed"
worm-like $X(s, t)$ swimmer
(e.g., spermatozoon)

- reduction of the number of model equations and
- avoiding the analysis of micro level interaction between a "solid" swimmer's body surface and fluid.

It should be noted along these lines that in typical swimming models dealing with "solid" swimmers, the latter are modeled as "traveling holes" in system's space domain, that is, the aforementioned "micro level" surface interaction is not in the picture as well.

On Well-Posedness of Swimming Models   To our knowledge, in the context of PDE approach to swimming modeling, the classical mathematical issues of well-posedness were addressed for the first time by Galdi [10] and [11] for a model of swimming micromotions in $\mathbb{R}^3$ (with the swimmer serving as the reference frame). In [48], San Martin et al. discussed the well-posedness of a 2D swimming model within the framework of the shape transformation approach for the fluid governed by the 2D Navier–Stokes equations.

On Models in the Context of ODEs   A number of attempts were made to introduce various reduction techniques to convert swimming model equations into systems of ODEs (e.g., by making use of empiric observations and experimental data, etc.), see, e.g., works by Becker et al. [2], Kanso et al. [18], Alouges et al. [1], Dal Maso et al. [6], and the references therein.

Part I
Modeling of Bio-Mimetic Swimmers in 2D and 3D Incompressible Fluids

Chapter 2
Bio-Mimetic Fish-Like Swimmers in a 2D Incompressible Fluid: Empiric Modeling

In this chapter we will describe a model of a bio-mimetic swimmer designed to imitate a swimming locomotion of a fish (or a sea snake) in a horizontal plane (see Figs. 1.1, 1.2 and 1.3). We will view this situation as the case of *a swimmer in a 2D incompressible fluid*. (Of course, a swimming motion in a horizontal plane in a 3D fluid is a much more complicated process, which we will also consider in Chap. 4.)

We will illustrate our modeling process below by making use of a swimmer, whose body consists of four identical rectangles as shown in Fig. 2.1.

2.1 Swimmer's Body as a Collection of Separate Sets

In this chapter and below, we will follow the modeling strategy outlined in Chap. 1, namely, given below:

- We assume that, at any moment of time t, our bio-mimetic swimmer is located within a given bounded fluid "swimming" domain (i.e., an open connected (measurable) set) $\Omega \subset R^K$, $K = 2, 3$, see more in Sect. 5.2.
- We identify the swimmer at time t with the fluid that lies within its body (regarded as a subset of Ω) at time t (see Remark 1.2).
- The body of our swimmer is a collection of n non-overlapping open connected (measurable) sets $S(z_i(t)) \subset \Omega \subset R^K$, $i = 1, \ldots, n$, $K = 2, 3$, where $z_i(t)$'s are the centers of mass of the respective sets $S(z_i(t))$'s at time t, see also Sect. 5.2 below for more formal assumptions.
- Every set $S(z_i(t))$, at any moment of time t, has a pre-described given shape, but its spatial orientation can vary in time, see also Sect. 5.2 below.
- These sets are connected by flexible links (i.e., they can change their length in time), which have a "negligible affect" on the swimming process and, therefore, such links are not explicitly present in our mathematical models below.

© The Author(s), under exclusive license to Springer Nature Switzerland AG 2021
A. Khapalov, *Bio-Mimetic Swimmers in Incompressible Fluids*, Lecture Notes in Mathematical Fluid Mechanics, https://doi.org/10.1007/978-3-030-85285-6_2

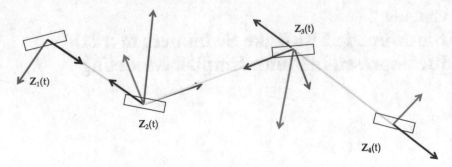

Fig. 2.1 A 2D fish- or sea snake-like bio-mimetic swimmer that consists of 4 rectangles with all internal forces engaged

Assumption 2.1 *Throughout this monograph, for simplicity of notations and to avoid some technicalities, we assume (unless we explicitly say otherwise) that*

1. *the fluid density $\rho = 1$;*
2. *the sets $S(z_i(t))$, $i = 1, \ldots, n$, are identical in mass, or, which is the same for $\rho = 1$, identical in the size of their measures (see Sect. 2.5.1 for otherwise).*

2.2   Bio-Mimetic Fish- and Snake-Like Swimmers

For this type of swimmers, we assume that the flexible links are available between the pairs $\{z_i(t), z_{(i-1)}(t)\}$, $i = 2, \ldots, n$, *only*.

Thus, at every moment of time, the swimmer's body is represented by a "broken-line" structure, formed by an ordered collection of points $z_1(t)), \ldots, z_n(t)$, which are surrounded by respective "supports" $S(z_i(t))$ (in Fig. 2.1 they are rectangles). We can say that the aforementioned points $z_i(t)$, $i = 1, \ldots, n$, form the skeleton of our bio-mimetic swimmer at time t.

2.3   Swimmer's Internal Forces

We assume that our bio-mimetic swimmer can engage two types of internal forces: rotational and elastic. These forces

- act through the immaterial links attached to the centers of mass $z_i(t)$'s of sets $S(z_i(t))$'s, and
- they create the same acceleration for all points within $S(z_i(t))$'s, attached to respective $z_i(t)$'s.

Thus, we can view and analyze the internal mechanics of the swimmer's body as a planar mechanics of a body consisting of n particles $z_i(t), i = 1, \ldots, n$, whose masses, due to Assumption 2.1 (see Sect. 2.5.1 for otherwise), are equal to the areas of their respective supports $S(z_i(t))$'s.

Mathematics-wise, we will model swimmer's forces acting upon each $S(z_i(t)), i = 1, \ldots, n$, as follows:

$$F_i(t, x) = F_{irot}(t, x) + F_{iel}(t, x), \quad i = 1, \ldots, n,$$

where

$$F_{irot}(t, x) = a_{irot}(t)\xi_i(t, x), \quad F_{iel}(t, x) = a_{iel}(t)\xi_i(t, x),$$

$a_{irot}(t)$ and $a_{iel}(t)$ are the vector accelerations, created by, respectively, rotational and elastic forces F_{irot} and F_{iel} at time t for all points within $S(z_i(t))$, and ξ_i's are the characteristic functions of $S(z_i(t))$'s:

$$\xi_i(t, x) = \begin{cases} 1, & \text{if } x \in S(z_i(t)), \\ 0, & \text{if } x \in \Omega \backslash S(z_i(t)), \end{cases} \quad i = 1, \ldots, n. \tag{2.3.1}$$

2.3.1 Rotational Internal Forces

In the case when the swimmer's body parts are identical in mass (see Assumption 2.1 and Sect. 2.5.1 for otherwise), the action of its internal rotational forces is described by the following formula:

$$F_{rot}(t, x) =$$

$$\sum_{i=2}^{n-1} v_{i-1}(t) \left[\xi_{i-1}(t, x) A(z_{i-1}(t) - z_i(t)) \right.$$

$$\left. -\xi_{i+1}(t, x) \frac{\|z_{i-1}(t) - z_i(t)\|^2}{\|z_{i+1}(t) - z_i(t)\|^2} A(z_{i+1}(t) - z_i(t)) \right]$$

$$+ \sum_{i=2}^{n-1} \xi_i(t, x) v_{i-1}(t) \left[A(z_i(t) \right.$$

$$\left. - z_{i-1}(t)) - \frac{\|z_{i-1}(t) - z_i(t)\|^2}{\|z_{i+1}(t) - z_i(t)\|^2} A(z_i(t) - z_{i+1}(t)) \right], \tag{2.3.2}$$

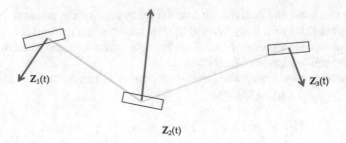

Fig. 2.2 A 2D swimmer consisting of 4 rectangles: only the rotational forces about $z_2(t)$ shown

where

$$A = \begin{pmatrix} 0 & 1 \\ -1 & 0 \end{pmatrix}.$$

Here, we assume the following:

- *Each of the "middle points" $z_i(t)$ for $i = 2, \ldots, n-1$ can force the two adjacent points to rotate about it.*
- All rotational forces act perpendicular to the respective flexible links.
- The magnitudes and directions of the rotational forces can be controlled via the selection of the coefficients $v_i(t)$, $i = 1, \ldots, n - 2$, which we regard as *multiplicative* controls (see [25]).
- The second sum on the right in (2.3.2) represents the counterforces generated by the pairs of rotational forces in the first sum, acting on the respective "middle points" according to Newton's Third Law, see Fig. 2.2 for illustration and Sect. 2.5.

2.3.2   Elastic Internal Forces

The shape of the swimmer is preserved, within "some limits" (in other words, our swimmer has a *flexible body*), by the elastic forces, which can act along the lines connecting *any pair of points* $\{z_i(t)\}$'s *that are connected by flexible links*, with the goal to preserve the distances between the swimmer's body parts as illustrated in Fig. 2.3. Such forces can act

1. according to Hooke's Law and/or
2. as the swimmer's *controlled* internal elastic forces.

Hooke's Elastic Forces In this case, we assume that, if the distances between any two adjacent points $z_i(t)$ and $z_{i-1}(t), i = 2, \ldots, n$, will deviate from the given initial value $l_{i-1} > 0, i = 2, \ldots, n$ (say, associated with "the state of rest" of the swimmer at hand), then an elastic force will act proportionally to the size of the

Fig. 2.3 A 2D swimmer consisting of 4 rectangles: only the elastic forces between $z_2(t)$ and $z_3(t)$ shown

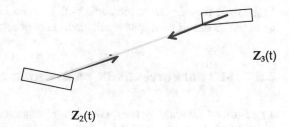

$Z_3(t)$

$Z_2(t)$

aforementioned deviation with coefficient $k_{i-1}, i = 2, \ldots, n$ with the goal to return the above distances to their initial values:

$$F_{hef}(t, x) = \sum_{i=2}^{n} [\xi_{i-1}(t, x) k_{i-1} \frac{(\|z_i(t) - z_{i-1}(t)\|_{R^2} - l_{i-1})}{\|z_i(t) - z_{i-1}(t)\|_{R^2}} (z_i(t) - z_{i-1}(t))$$

$$+ \xi_i(t, x) k_{i-1} \frac{(\|z_i(t) - z_{i-1}(t)\|_{R^2} - l_{i-1})}{\|z_i(t) - z_{i-1}(t)\|_{R^2}} (z_{i-1}(t) - z_i(t))]. \qquad (2.3.3)$$

Controlled Internal Elastic Forces In this case we assume that the swimmer can vary the size of k_i's in the above. For simplicity of representation, we will use *controls* $w_i(t), i = 1, \ldots, n-1$, and eliminate the ratios in (2.3.3):

$$(F_{cef}(t, x) = \sum_{i=2}^{n} [\xi_{i-1}(t, x) w_{i-1}(t)(z_i(t) - z_{i-1}(t)) + \xi_i(t, x) w_{i-1}(t)(z_{i-1}(t) - z_i(t))].$$

$$(2.3.4)$$

Remark 2.1 (On Well-Posedness of the Integral Structure of Swimmer's Body) The structural integrity of swimmer's body is one of the goals for a controlled swimmer's locomotion in this monograph. In our results below, we intend to avoid the situation when the sets $S(z_i(t))$'s overlap ("collide") or when the distance between them would become "unreasonably small or large." In the real world, such situations are possible, for example, a body of a ship can be destroyed in a storm by massive waves, etc.

2.4 Swimmer's Geometric Controls

Throughout this monograph, we assume that each part of the swimmer's body at hand is a given pre-described set (a swimmer can have parts of different pre-described shape, see also Assumption 2.1), which can change its spatial orientation, relative to its center of mass, in time. In Introduction, we regarded this aspect of swimmers' possibilities as the **geometric controls**.

For example, in the case of the swimmer in Fig. 2.1, all its body parts have a pre-described shape of the same rectangle, say, $(a, b) \times (c, d)$, while any of the sets

$S(z_i(t))$'s is the very same rectangle with its center located at point $z_i(t)$, and each of $S(z_i(t))$'s has its own spatial orientation about $z_i(t)$ at time t.

2.5   Internal Forces and Conservation of Momenta

The Sum of Internal Forces We want to emphasize again that the sum of all the swimmer's forces in (2.3.2)–(2.3.4) is equal to zero. Thus, these forces cannot move its center of mass without interaction with a surrounding medium. This is a principal characteristic of an object which is "swimming-by-itself" in a fluid.

Conservation of Momenta In our calculations below, we regard (as we mentioned it in the beginning of Sect. 2.3) the swimmer's internal forces as the forces that act on n particles $z_i(t), i = 1, \ldots, n$, whose masses are equal to the areas of their respective supports $S(z_i(t))$'s. All these areas are equal to one another due to Assumption 2.1, and, say to meas$(S) = $ meas$(S(z_i(t))), i = 1, \ldots, n$, where "meas" stands for "measure," and, e.g., $S = S(z_1(0))$ (see also Assumption (A5.1) in Chap. 5 below about our standard selection for S).

Let us evaluate the torques created by the swimmer's internal forces:

- Since the pairs of counteracting (due to Newton's Third Law) elastic forces in (2.3.3)–(2.3.4) act along the lines connecting the respective $z_i(t)$'s, parallel to each other, they create the zero torque.
- We will now show that the torque, created by the swimmer's rotational forces in (2.3.2), is also equal to zero. Let us compute the torque T_2, generated by the pair of rotational forces about $z_2(t)$.

  Making use of (3.45) on p. 83 in [53], we obtain

$$T_2 = -\|v_1(t)A(z_1(t) - z_2(t))\|_{R^2}\|z_1(t) - z_2(t)\|_{R^2} \text{ (meas }(S))$$

$$+ \left\| v_1(t)\frac{\|z_1(t) - z_2(t)\|_{R^2}^2}{\|z_3(t) - z_2(t)\|_{R^2}^2}(-A)(z_3(t) - z_2(t)) \right\|_{R^2} \|z_3(t) - z_2(t)\|_{R^2} \text{ (meas }(S))$$

$$= -|v_1(t)|\|z_1(t) - z_2(t)\|_{R^2}^2 + |v_1(t)|\|z_1(t) - z_2(t)\|_{R^2}^2 \text{ (meas }(S)) = 0.$$

We can compute the torques $T_3, \ldots, T_{n-1}$ generated by $z_3(t), \ldots, z_{n-1}(t)$ in a similar manner. Thus, the angular momentum of our swimmer is conserved.

2.5.1   About Swimmers with Body Parts Different in Mass

In the case when the parts of the swimmer at hand have different masses (i.e., have different areas of support), the formulas (2.3.2)–(2.3.4) should be proportionally adjusted to ensure that the torque created by the swimmer's rotational forces is equal to zero and the sum of all its internal forces is also equal to zero. Namely, in this case, we would have the following formula for the combined swimmer's internal forces:

$$F_{rot}^*(t, x) = \sum_{i=2}^{n-1} v_{i-1}(t)[\xi_{i-1}(t, x)A(z_{i-1}(t) - z_i(t))$$

$$-\xi_{i+1}(t, x)\frac{\|z_{i-1}(t) - z_i(t)\|^2}{\|z_{i+1}(t) - z_i(t)\|^2}A(z_{i+1}(t) - z_i(t))\frac{(\text{meas}(S(z_{i-1}(t)))}{(\text{meas}(S(z_{i+1}(t)))}]$$

$$+\sum_{i=2}^{n-1} \xi_i(t, x)v_{i-1}(t)[A(z_i(t) - z_{i-1}(t))\frac{(\text{meas}(S(z_{i-1}(t)))}{(\text{meas}(S(z_i(t)))}$$

$$-\frac{\|z_{i-1}(t) - z_i(t)\|^2}{\|z_{i+1}(t) - z_i(t)\|^2}A(z_i(t) - z_{i+1}(t))\frac{(\text{meas}(S(z_{i-1}(t)))}{(\text{meas}(S(z_{i+1}(t)))}\frac{(\text{meas}(S(z_{i+1}(t)))}{(\text{meas}(S(z_i(t)))}],$$

$$F_{hef}^*(t, x) = \sum_{i=2}^{n}[\xi_{i-1}(t, x)k_{i-1}\frac{(\|z_i(t) - z_{i-1}(t)\|_{R^2} - l_{i-1})}{\|z_i(t) - z_{i-1}(t)\|_{R^2}}(z_i(t) - z_{i-1}(t))$$

$$+\xi_i(t, x)k_{i-1}\frac{(\|z_i(t) - z_{i-1}(t)\|_{R^2} - l_{i-1})}{\|z_i(t) - z_{i-1}(t)\|_{R^2}}(z_{i-1}(t) - z_i(t))\frac{(\text{meas}(S(z_{i-1}(t)))}{(\text{meas}(S(z_i(t)))}],$$

$$F_{cef}^*(t, x) =$$

$$\sum_{i=2}^{n}[\xi_{i-1}(x, t)w_{i-1}(z_i(t) - z_{i-1}(t)) + \xi_i(x, t)w_{i-1}(z_{i-1}(t) - z_i(t))\frac{(\text{meas}(S(z_{i-1}(t)))}{(\text{meas}(S(z_i(t)))}.$$

2.6 Fluid Equations: Non-stationary Stokes and Navier–Stokes Equations in 2D

The initial- and boundary-value problem for the non-stationary Stokes equations looks as follows:

$$\begin{cases} u_t - \nu\Delta u + \frac{1}{\rho}\nabla p = \frac{1}{\rho}f & \text{in } Q_T = (0, T) \times \Omega, \\ \text{div } u = 0 & \text{in } Q_T, \\ u = 0 & \text{in } \Sigma_T = (0, T) \times \partial\Omega, \\ u(0, \cdot) = u_0 & \text{in } \Omega. \end{cases} \quad (2.6.1)$$

Here, $x = (x_1, x_2)$, $\partial\Omega$ is a boundary of $\Omega \subset R^2$, $u(t, x)$ and $p(t, x)$ are, respectively, the velocity of the fluid and its pressure at point x at time t,

$$u(t, x) = (u_1(t, x), u_2(t, x)),$$

f is a force term,

$$f(t, x) = (f_1(t, x), f_2(t, x)),$$

$$\text{div } u = u_{1x_1} + u_{2x_2} = 0$$

is the incompressibility assumption, $u_0(x)$ is the initial velocity, ρ is the fluid density (see Assumption 2.1), and ν is the kinematic viscosity constant.

In the case of the 2D Navier–Stokes equations, *a nonlinear "inertia" term* is added to the above non-stationary Stokes equations:

$$\begin{cases} u_t - \nu\Delta u + (u \cdot \nabla)u + \frac{1}{\rho}\nabla p = \frac{1}{\rho}f & \text{in } Q_T = (0, T) \times \Omega, \\ \text{div } u = 0 & \text{in } Q_T, \\ u = 0 & \text{in } \Sigma_T = (0, T) \times \partial\Omega, \\ u(0, \cdot) = u_0 & \text{in } \Omega, \end{cases} \tag{2.6.2}$$

Remark 2.2 (Stationary Stokes Equations vs. Non-stationary Stokes Equations) In various publications discussed in the bibliography Sect. 1.4 of Introduction, a frequent choice of fluid for microswimmers is the fluid governed by the *stationary* Stokes equation. The empiric reasoning behind this is that, due to the small size of swimmer, the inertia terms in the Navier–Stokes equation, containing the first-order derivatives in t and x, can be omitted, provided that the respective frequency parameter is of order one. However, it was noted that a microswimmer (e.g., a nano-size robot) may use a rather high frequency of motion, which may justify at least in some cases the presence for the term u_t in the model.

Furthermore, the models for swimmers in a fluid described by the non-stationary Stokes equations seem to us to be a natural step toward the models in a fluid governed by the Navier–Stokes equations (as it is confirmed by technical calculations below in this monograph).

2.7 A Model of a 2D Fish-Like Bio-Mimetic Swimmer: The Case of Stokes Equations

Under Assumption 2.1, combining the above, we obtain the following model of our fish-like artificial swimmer in a fluid described by the 2D non-stationary Stokes equations with the controlled elastic forces $F_{cef}(t, x)$ only (we can, of course, consider any combination of $F_{hef}(t, x)$ and $F_{cef}(t, x)$ forces from Sect. 2.3.2):

$$
\begin{cases}
u_t = \nu \Delta u + F_{fish2D} - \nabla p & \text{in } Q_T = (0, T) \times \Omega, \\
\text{div } u = 0 & \text{in } Q_T, \\
u = 0 & \text{in } \Sigma_T = (0, T) \times \partial\Omega, \\
u(0, \cdot) = u_0 & \text{in } \Omega \subset R^2, \\
\dfrac{dz_i}{dt} = \dfrac{1}{\text{meas}(S)} \int_{S(z_i(t))} u(t, x) dx, \quad z_i(0) = z_{i,0}, \quad i = 1, \dots, n,
\end{cases}
$$

$$(2.7.1)$$

where

$$
z(t) = (z_1(t), \dots, z_n(t)) \in [R^2]^n, \quad v(t) = (v_1(t), \dots, v_{n-2}(t)) \in R^{n-2},
$$

$$
w(t) = (w_1(t), \dots, w_{n-1}(t)) \in R^{n-1},
$$

$$
F_{fish2D}(t, x) = \sum_{i=2}^{n-1} v_{i-1}(t)[\xi_{i-1}(t, x) A(z_{i-1}(t) - z_i(t))
$$

$$
-\xi_{i+1}(t, x) \frac{\|z_{i-1}(t) - z_i(t)\|^2}{\|z_{i+1}(t) - z_i(t)\|^2} A(z_{i+1}(t) - z_i(t))]
$$

$$
+ \sum_{i=2}^{n-1} \xi_i(t, x) v_{i-1}(t) \left[A(z_i(t) - z_{i-1}(t)) - \frac{\|z_{i-1}(t) - z_i(t)\|^2}{\|z_{l+1}(t) \quad z_i(t)\|^2} A(z_i(t) - z_{i+1}(t)) \right]
$$

$$
+ \sum_{i=2}^{n} [\xi_{i-1}(t, x) w_{i-1}(z_i(t) - z_{i-1}(t)) + \xi_i(t, x) w_{i-1}(z_{i-1}(t) - z_i(t))]. \quad (2.7.2)
$$

In the above *hybrid coupled system of fluid equations and a system of integro-ordinary differential equations* (2.7.2),

- 1. the former describes the evolution of an incompressible fluid governed by the 2D non-stationary Stokes equations with a given fluid velocity u_0 that evolves and
  2. under the influence of the forcing term $F(z)$ that represents the actions of the swimmer at hand, in the form of explicit expressions (2.3.2)–(2.3.4), on the fluid surrounding it.
- In turn, the latter describes the motion of the swimmer's body parts within Ω through the motion of their respective centers of mass $z_i(t)$'s, which allows one to calculate the motion of the swimmer's center of mass $z_c(t)$; namely,

$$
z_c(t) = \frac{1}{n} \sum_{i=1}^{n} z_i(t). \quad (2.7.3)
$$

Remark 2.3 Throughout the monograph, we denote the (same) force terms in the fluid equations either by symbol F or by $F(z)$ (respectively, $F(t, x) = (F(z))(t, x)$), in some cases with subscripts. The latter notation is used to remind the reader, in some contexts, that the swimmer's internal forces depend on the positions $z_i(t), i = 1, \ldots, n$ (see also (2.3.1) in this regard).

2.8 A Model of a 2D Fish-Like Bio-Mimetic Swimmer: The Case of Navier–Stokes Equations

Again, under Assumption 2.1, in the notations of Sect. 2.7, we obtain the following model of our fish-like artificial swimmer in a fluid described by the 2D Navier–Stokes equations:

$$\begin{cases} u_t + (u \cdot \nabla)u = v \Delta u + F_{fish2D} - \nabla p & \text{in } Q_T = (0, T) \times \Omega, \\ \text{div } u = 0 & \text{in } Q_T, \\ u = 0 & \text{in } \Sigma_T = (0, T) \times \partial \Omega, \\ u(0, \cdot) = u_0 & \text{in } \Omega \subset R^2, \\ \frac{dz_i}{dt} = \frac{1}{\text{meas}(S)} \int_{S(z_i(t))} u(t, x) dx, \quad z_i(0) = z_{i,0}, \quad i = 1, \ldots, n, \end{cases}$$

$$(2.8.1)$$

where, similar to (2.7.1), $F_{fish2D}(z)$ includes the rotational forces and any combination of $F_{hef}(t, x)$ and $F_{cef}(t, x)$ forces from Sect. 2.3.2.

Chapter 3
Bio-Mimetic Aquatic Frog- and Clam-Like Swimmers in a 2D Fluid: Empiric Modeling

In this chapter, based on the general formulas of Chap. 2, we will describe two models of bio-mimetic swimmers designed to imitate swimming patterns of a frog and a clam in a horizontal plane. Once again, we will view this situation as the case of "swimmers in a 2D incompressible fluid."

Remark 3.1 (Other 2D (or 3D) Models) In general, one can consider infinitely many other 2D (or 3D, as in Chap. 4) models of swimmers created from available separate parts along the guidelines in Chap. 2. In this monograph, we chose to focus on three real-world patterns of swimming locomotion—*of a fish, a frog, and a clam*—as observed in nature. However, most of the mathematical results below in Parts II–V apply to any model of swimmers constructed along the modeling strategy described in Chaps. 1 and 2.

3.1 A Bio-Mimetic Aquatic Frog-Like Swimmer in a 2D Incompressible Fluid

In this case we will consider the model of a bio-mimetic swimmer in Fig. 3.2 proposed here to imitate the mechanics of a frog, see Fig. 3.1. While in Fig. 3.2, the parts of the swimmer's body are rectangles, in the general case they are assumed to be as in Assumption 2.1.

Similar to the derivation of model (2.7.1)–(2.7.2) in Sect. 2.7, again under Assumptions 2.1, we can obtain the following model for our frog-like artificial swimmer in a fluid described by the 2D non-stationary Stokes equations with

Fig. 3.1 An aquatic frog

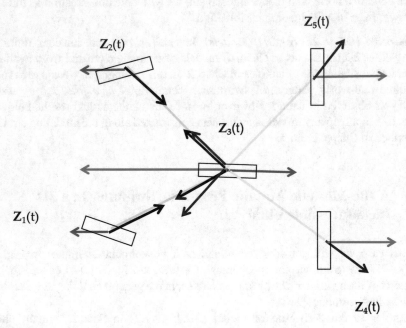

Fig. 3.2 An aquatic 2D frog-like bio-mimetic swimmer that consists of 5 rectangles with all internal forces engaged

controlled elastic forces (Hooke's elastic forces can also be considered, of course):

$$
\begin{cases}
u_t = \nu \Delta u + F_{frog2D} - \nabla p & \text{in } Q_T = (0, T) \times \Omega, \\
\text{div } u = 0 & \text{in } Q_T, \\
u = 0 & \text{in } \Sigma_T = (0, T) \times \partial\Omega, \\
u(0, \cdot) = u_0 & \text{in } \Omega \subset R^2, \\
\frac{dz_i}{dt} = \frac{1}{\text{meas}(S)} \int_{S(z_i(t))} u(t, x) dx, \quad z_i(0) = z_{i,0}, \quad i = 1, \ldots, 5,
\end{cases}
$$

(3.1.1)

where

$$
z(t) = (z_1(t), \ldots, z_5(t)) \in [R^2]^5, \quad v(t) = (v_1(t), v_2(t)), \quad w(t) = (w_1(t), \ldots, w_4(t)),
$$

$$
F_{frog2D}(t, x) = v_1(t)[\xi_1(t, x) A(z_1(t) - z_3(t))
$$

$$
- \xi_2(t, x) \frac{\|z_1(t) - z_3(t)\|^2}{\|z_2(t) - z_3(t)\|^2} A(z_2(t) - z_3(t))]
$$

$$
+ v_2(t) \left[\xi_4(t, x) A(z_4(t) - z_3(t)) - \xi_2(t, x) \frac{\|z_4(t) - z_3(t)\|^2}{\|z_5(t) - z_3(t)\|^2} A(z_5(t) - z_3(t)) \right]
$$

$$
+ \xi_3(t, x) v_1(t) \left[A(z_3(t) - z_1(t)) - \frac{\|z_1(t) - z_3(t)\|^2}{\|z_2(t) - z_3(t)\|^2} A(z_3(t) - z_2(t)) \right]
$$

$$
+ \xi_3(t, x) v_2(t) \left[A(z_3(t) - z_4(t)) - \frac{\|z_4(t) - z_3(t)\|^2}{\|z_5(t) - z_3(t)\|^2} A(z_3(t) - z_5(t)) \right]
$$

$$
+ \xi_1(t, x) w_1(z_3(t) - z_1(t)) + \xi_3(t, x) w_1(z_1(t) - z_3(t))
$$

$$
+ \xi_2(t, x) w_2(z_3(t) - z_2(t)) + \xi_3(t, x) w_2(z_2(t) - z_3(t))
$$

$$
+ \xi_4(t, x) w_3(z_3(t) - z_4(t)) + \xi_3(t, x) w_3(z_4(t) - z_3(t))
$$

$$
+ \xi_5(t, x) w_4(z_3(t) - z_5(t)) + \xi_3(t, x) w_4(z_5(t) - z_3(t)).
$$

(3.1.2)

In the case of the Navier–Stokes equations, we need to replace (3.1.1) with the following hybrid coupled system:

$$
\begin{cases}
u_t + (u \cdot \nabla)u = \nu \Delta u + F_{frog2D} - \nabla p & \text{in } Q_T = (0, T) \times \Omega, \\
\text{div } u = 0 & \text{in } Q_T, \\
u = 0 & \text{in } \Sigma_T = (0, T) \times \partial\Omega, \\
u(0, \cdot) = u_0 & \text{in } \Omega \subset R^2, \\
\frac{dz_i}{dt} = \frac{1}{\text{meas}(S)} \int_{S(z_i(t))} u(t, x) dx, \quad z_i(0) = z_{i,0}, \quad i = 1, \ldots, 5.
\end{cases}
$$

(3.1.3)

3.2   A Bio-Mimetic Clam-Like Swimmer in a 2D Incompressible Fluid

In this case we will consider the model of a bio-mimetic swimmer in Fig. 3.5, proposed here to imitate the mechanics of a swimming clam (file clam or scallop), see Figs. 3.3 and 3.4. In Fig. 3.5, the parts of the swimmer's body are rectangles, in the general case they are assumed to be as in Assumption 2.1.

In fact, the "imitation" here is very far away from being a "true imitation," namely,

- an actual clam like in Fig. 3.3 or Fig. 3.4 has only one joint and it can only open and close two walls of its shell, while
- our "bio-mimetic clam" in Fig. 3.5 has four joints:

  1. one is due to the rotational forces about the point $z_1(t)$ and
  2. three others are due to the geometric controls that allow our artificial swimmer to change the spatial orientations of all 3 rectangles about their respective centers of mass $z_1(t)$, $z_2(t)$, and $z_3(t)$.

- The clam in nature seems to be capable of employing some (water) jet propulsion technique which allows it to move in the direction of the opening.

Similar to the derivation of models (2.7.1)–(2.7.2) in Sect. 2.7 and (3.1.1)–(3.1.2) in the above section, again under Assumptions 2.1, we can obtain the following model for our clam-like artificial swimmer in a fluid described by the 2D non-stationary Stokes equations with *controlled elastic forces* (Hooke's elastic forces

Fig. 3.3 A file clam

Fig. 3.4 A scallop

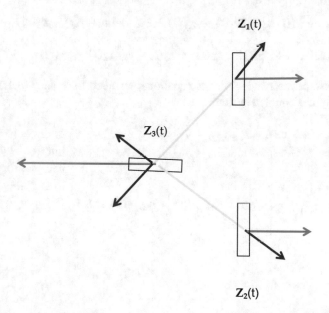

Fig. 3.5 A 2D clam-like bio-mimetic swimmer that consists of 3 rectangles with all internal forces engaged

can also be considered, of course):

$$\begin{cases} u_t = \nu \Delta u + F_{clam2D} - \nabla p & \text{in } Q_T = (0, T) \times \Omega, \\ \text{div } u = 0 & \text{in } Q_T, \\ u = 0 & \text{in } \Sigma_T = (0, T) \times \partial \Omega, \\ u(0, \cdot) = u_0 & \text{in } \Omega \subset R^2, \\ \frac{dz_i}{dt} = \frac{1}{\text{meas}(S)} \int_{S(z_i(t))} u(t, x) dx, \quad z_i(0) = z_{i,0}, \quad i = 1, 2, 3, \end{cases}$$

(3.2.1)

where

$$z(t) = (z_1(t), z_2(t), z_3(t)) \in [R^2]^3, \quad w(t) = (w_1(t), w_2(t)),$$

$$F_{clam2D}(t, x) = v_1(t)[\xi_1(t, x)A(z_1(t) - z_3(t))$$

$$-\xi_2(t, x)\frac{\|z_1(t) - z_3(t)\|^2}{\|z_2(t) - z_3(t)\|^2}A(z_2(t) - z_3(t))]$$

$$+ \xi_3(t, x)v_1(t)\left[A(z_3(t) - z_1(t)) - \frac{\|z_1(t) - z_3(t)\|^2}{\|z_2(t) - z_3(t)\|^2}A(z_3(t) - z_2(t))\right]$$

$$+ \xi_1(t, x)w_1(z_3(t) - z_1(t)) + \xi_3(t, x)w_1(z_1(t) - z_3(t))$$

$$+ \xi_2(t, x)w_2(z_3(t) - z_2(t)) + \xi_3(t, x)w_2(z_2(t) - z_3(t)).$$

(3.2.2)

In the case of the Navier–Stokes equations, we need to replace (3.1.1) with the following hybrid coupled system:

$$\begin{cases} u_t + (u \cdot \nabla)u = \nu \Delta u + F_{clam2D} - \nabla p & \text{in } Q_T = (0, T) \times \Omega, \\ \text{div } u = 0 & \text{in } Q_T, \\ u = 0 & \text{in } \Sigma_T = (0, T) \times \partial \Omega, \\ u(0, \cdot) = u_0 & \text{in } \Omega \subset R^2, \\ \frac{dz_i}{dt} = \frac{1}{\text{meas}(S)} \int_{S(z_i(t))} u(t, x) dx, \quad z_i(0) = z_{i,0}, \quad i = 1, 2, 3. \end{cases}$$

(3.2.3)

Chapter 4
Bio-Mimetic Swimmers in a 3D Incompressible Fluid: Empiric Modeling

In this chapter, we extend the 2D models of bio-mimetic fish-, frog-, and clam-like swimmers to the 3D case.

4.1 Rotational Forces in 3D

As in Sect. 2.3.1 of Chap. 2, we assume that

- each of the "middle" points $z_i(t)$'s, *among those that are capable of engaging rotational forces*, can force the two adjacent points $\{z_{i-1}(t), z_{i+1}(t)\}$ to rotate about it;
- all rotational forces act perpendicular to the respective flexible links.

However, in 3D we will add a **new assumption**, which will allow us to conserve the angular momentum of the swimmer at hand (see Sect. 2.5), namely,

- for each triplet of points $\{z_{i-1}(t), z_i(t), z_{i+1}(t)\}$ that participate in the action of rotational forces about $z_i(t)$, both the rotational forces, applied to $\{z_{i-1}(t), z_{i+1}(t)\}$, and their counterforce, applied to $z_i(t)$, lie in the **same plane** at every moment of time.

Let us elaborate on this new assumption.

In the two-dimensional space, all the forces lie in the same plane at all times, and we can describe them by making use of the matrix

$$A = \begin{pmatrix} 0 & 1 \\ -1 & 0 \end{pmatrix}$$

as it is given in (2.3.2) in Sect. 2.3.1 of Chap. 2.

In the three-dimensional space, to satisfy Newton's Third Law and to conserve the angular momentum of swimmers' internal forces, we also need to make sure that the respective rotational forces, acting on $z_{i-1}(t)$ and $z_{i+1}(t)$, and their counterforce, acting on $z_i(t)$, lie in one plane, namely, spanned by the vectors $z_{i-1}(t) - z_i(t)$ and $z_{i+1}(t) - z_i(t)$, respectively.

Continuity of Transformation of the Plane for the Rotational Forces If, at some moment of time t_*, vectors $z_{i-1}(t) - z_i(t)$ and $z_{i+1}(t) - z_i(t)$ are co-linear, the respective links connecting $z_i(t)$ with $z_{i-1}(t)$ and $z_{i+1}(t)$ lie in an aligned configuration. In such a configuration, we assume that a swimmer can *instantaneously* choose an *arbitrary* new rotational plane for the triplet of points $\{z_{i-1}(t), z_i(t), z_{i+1}(t)\}$ for $t > t_*$, among infinitely many available planes.

Thus, the above-described aligned configuration is an *intrinsic* point of "*geometric*" *discontinuity* of the rotational forces at time $t = t_*$, if the magnitudes of the respective rotational forces are *strictly separated from zero near* $t = t_*$.

To overcome this difficulty, we define the 3D rotational forces to be vanishing at the aforementioned aligned configurations, namely, as follows (compare to (2.3.2)):

$$
\begin{aligned}
F_{rot3D}(t, x) := & \sum_{i=2}^{N-1} v_{i-1}(t) \bigg[\xi_{i-1}(t, x) \ A_i(t) \big(z_{i-1}(t) - z_i(t) \big) \\
& - \xi_{i+1}(t, x) \frac{|z_{i-1}(t) - z_i(t)|^2}{|z_{i+1}(t) - z_i(t)|^2} \ B_i(t) \big(z_{i+1}(t) - z_i(t) \big) \bigg] \\
& + \sum_{i=2}^{N-1} \xi_i(t, x) \, v_{i-1}(t) \bigg[A_i(t) \big(z_i(t) - z_{i-1}(t) \big) \\
& - \frac{|z_{i-1}(t) - z_i(t)|^2}{|z_{i+1}(t) - z_i(t)|^2} \ B_i(t) \big(z_i(t) - z_{i+1}(t) \big) \bigg],
\end{aligned} \tag{4.1.1}
$$

where the scalar functions $v_1(t), \ldots, v_{N-2}(t)$ control the magnitudes of the rotational forces and determine whether they act in folding or unfolding fashion, and

$$
x \mapsto A_i(t)x := \big[(z_{i-1}(t) - z_i(t)) \times (z_{i+1}(t) - z_i(t)) \big] \times x,
$$

$$
x \mapsto B_i(t)x := x \times \big[(z_{i-1}(t) - z_i(t)) \times (z_{i+1}(t) - z_i(t)) \big], \quad x = (x_1, x_2, x_3).
$$

Note that $A_i(t)x = -B_i(t)x$ and $|A_i(t)\mathbf{x}| = |B_i(t)x| \to 0$ for any x when points $z_{i-1}(t), z_i(t)$, and $z_{i+1}(t)$ converge to an aligned configuration (Figs. 4.1 and 4.2).

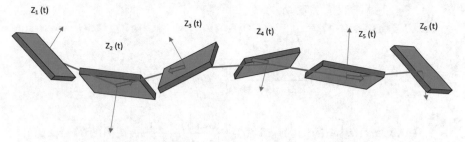

Fig. 4.1 A 3D bio-mimetic "fish-like" swimmer whose body consists of 6 identical parallelepipeds with some internal structural and rotational forces engaged: position I

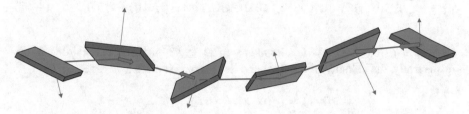

Fig. 4.2 A 3D bio-mimetic "fish-like" swimmer whose body consists of 6 identical parallelepipeds: position II

4.2   A Model of a 3D Fish-Like Bio-Mimetic Swimmer: The Case of Stokes Equations

Under Assumption 2.1, combining the above, we obtain the following model for our fish-like artificial swimmer in a fluid described by the 3D non-stationary Stokes equations with the controlled elastic forces $F_{cef}(t, x)$ only (we can, of course, consider any combination of Hooke's elastic and controlled elastic forces from Sect. 2.3.2):

$$\begin{cases} u_t = \nu \Delta u + F_{fish3D} - \nabla p & \text{in } Q_T = (0, T) \times \Omega, \\ \text{div } u = 0 & \text{in } Q_T, \\ u = 0 & \text{in } \Sigma_T = (0, T) \times \partial \Omega, \\ u(0, \cdot) = u_0 & \text{in } \Omega \subset R^3, \\ \frac{dz_i}{dt} = \frac{1}{\text{meas}(S)} \int_{S(z_i(t))} u(t, x) dx, \quad z_i(0) = z_{i,0}, \quad i = 1, \ldots, n, \end{cases}$$

(4.2.1)

where

$$z(t) = (z_1(t), \ldots, z_n(t)) \in [R^3]^n, \quad v(t) = (v_1(t), \ldots, v_{n-2}(t)) \in R^{n-2},$$

$$w(t) = (w_1(t), \ldots, w_{n-1}(t)) \in R^{n-1},$$

$$F_{fish3D}(t, x) = \sum_{i=2}^{N-1} v_{i-1}(t)\left[\xi_{i-1}(t, x) \; A_i(t)\big(z_{i-1}(t) - z_i(t)\big)\right.$$

$$\left. -\xi_{i+1}(t, x) \frac{|z_{i-1}(t) - z_i(t)|^2}{|z_{i+1}(t) - z_i(t)|^2} \; B_i(t)\big(z_{i+1}(t) - z_i(t)\big)\right]$$

$$+ \sum_{i=2}^{N-1} \xi_i(t, x) v_{i-1}(t)\left[A_i(t)\big(z_i(t) - z_{i-1}(t)\big) - \frac{|z_{i-1}(t) - z_i(t)|^2}{|z_{i+1}(t) - z_i(t)|^2} \; B_i(t)\big(z_i(t) - z_{i+1}(t)\big)\right],$$

$$+ \sum_{i=2}^{n} [\xi_{i-1}(t, x) w_{i-1}(z_i(t) - z_{i-1}(t)) + \xi_i(t, x) w_{i-1}(z_{i-1}(t) - z_i(t))]. \tag{4.2.2}$$

Here, $x = (x_1, x_2, x_3)$, $\partial\Omega$ is a boundary of $\Omega \subset R^3$, and $u(t, x)$ and $p(t, x)$ are, respectively, the velocity of the fluid and its pressure at point x at time t,

$$u(t, x) = (u_1(t, x), u_2(t, x), u_3(t, x)).$$

Again, under Assumption 2.1, we obtain the following model of our fish-like artificial swimmer in a fluid described by the 3D Navier–Stokes equations:

$$\begin{cases} u_t + (u \cdot \nabla)u = v\Delta u + F_{fish3D} - \nabla p & \text{in } Q_T = (0, T) \times \Omega, \\ \text{div } u = 0 & \text{in } Q_T, \\ u = 0 & \text{in } \Sigma_T = (0, T) \times \partial\Omega, \\ u(0, \cdot) = u_0 & \text{in } \Omega \subset R^3, \\ \frac{dz_i}{dt} = \frac{1}{\text{meas}(S)} \int_{S(z_i(t))} u(t, x)dx, \; z_i(0) = z_{i,0}, \; i = 1, \ldots, n, \end{cases}$$

$$\tag{4.2.3}$$

where, similar to (4.2.1), $F_{fish3D}(z)$ includes the rotational forces and any combination of Hooke's elastic and controlled elastic forces as in Sect. 2.3.2.

4.3 A Bio-Mimetic Frog-Like Swimmer in a 3D Incompressible Fluid

In this case we will consider a model of a bio-mimetic swimmer as in Fig. 4.3. While in Fig. 4.3 the parts of the swimmer's body are 5 identical parallelepipeds, in the general case they are assumed to be as described in Assumption 2.1.

Similar to the derivation of model (3.1.1)–(3.1.2) in Sect. 3.1, again under Assumption 2.1, we can obtain the following model for our frog-like artificial

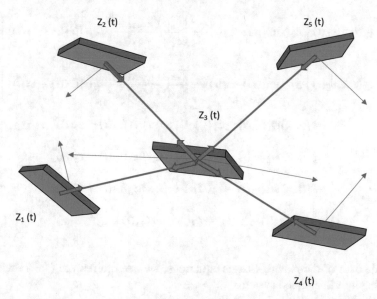

Fig. 4.3 A 3D frog-like bio-mimetic swimmer that consists of 5 identical parallelepipeds with all internal forces engaged

swimmer in a fluid described by the 3D non-stationary Stokes equations with *controlled elastic forces* (Hooke's elastic forces can also be considered, of course):

$$
\begin{cases}
u_t = v\Delta u + F_{frog3D} - \nabla p & \text{in } Q_T = (0, T) \times \Omega, \\
\operatorname{div} u = 0 & \text{in } Q_T, \\
u = 0 & \text{in } \Sigma_T = (0, T) \times \partial\Omega, \\
u(0, \cdot) = u_0 & \text{in } \Omega \subset R^3, \\
\dfrac{dz_i}{dt} = \dfrac{1}{\operatorname{meas}(S)} \int_{S(z_i(t))} u(t, x)dx, \quad z_i(0) = z_{i,0}, \quad i = 1, \ldots, 5,
\end{cases}
$$

$$(4.3.1)$$

where

$$
z(t) = (z_1(t), \ldots, z_5(t)) \in [R^2]^5, \quad v(t) = (v_1(t), v_2(t)), \quad w(t) = (w_1(t), \ldots, w_4(t)),
$$

$$
F_{frog3D}(t, x) = v_1(t)\left[\xi_1(t, x)A_i(t)(z_1(t) - z_3(t))\right.
$$

$$
\left. -\xi_2(t, x)\frac{\|z_1(t) - z_3(t)\|^2}{\|z_2(t) - z_3(t)\|^2}B_i(t)(z_2(t) - z_3(t))\right]
$$

$$
+ v_2(t)\left[\xi_4(t, x)A_i(t)(z_4(t) - z_3(t)) - \xi_2(t, x)\frac{\|z_4(t) - z_3(t)\|^2}{\|z_5(t) - z_3(t)\|^2}B_i(t)(z_5(t) - z_3(t))\right]
$$

$$+ \xi_3(t, x)v_1(t) \left[A_i(t)(z_3(t) - z_1(t)) - \frac{\|z_1(t) - z_3(t)\|^2}{\|z_2(t) - z_3(t)\|^2} B_i(t)(z_3(t) - z_2(t)) \right]$$

$$+ \xi_3(t, x)v_2(t) \left[A_i(t)(z_3(t) - z_4(t)) - \frac{\|z_4(t) - z_3(t)\|^2}{\|z_5(t) - z_3(t)\|^2} B_i(t)(z_3(t) - z_5(t)) \right]$$

$$+ \xi_1(t, x)w_1(z_3(t) - z_1(t)) + \xi_3(t, x)w_1(z_1(t) - z_3(t))$$

$$+ \xi_2(t, x)w_2(z_3(t) - z_2(t)) + \xi_3(t, x)w_2(z_2(t) - z_3(t))$$

$$+ \xi_4(t, x)w_3(z_3(t) - z_4(t)) + \xi_3(t, x)w_3(z_4(t) - z_3(t))$$

$$+ \xi_5(t, x)w_4(z_3(t) - z_5(t)) + \xi_3(t, x)w_4(z_5(t) - z_3(t)). \tag{4.3.2}$$

In the case of the Navier–Stokes equations, we need to replace (4.3.1) with the following hybrid coupled system:

$$\begin{cases} u_t + (u \cdot \nabla)u = \nu \Delta u + F_{frog} - \nabla p & \text{in } Q_T = (0, T) \times \Omega, \\ \operatorname{div} u = 0 & \text{in } Q_T, \\ u = 0 & \text{in } \Sigma_T = (0, T) \times \partial\Omega, \\ u(0, \cdot) = u_0 & \text{in } \Omega \subset R^3, \\ \frac{dz_i}{dt} = \frac{1}{\operatorname{meas}(S)} \int_{S(z_i(t))} u(t, x)dx, \quad z_i(0) = z_{i,0}, \quad i = 1, \ldots, 5. \end{cases}$$

$$\tag{4.3.3}$$

4.4   A Bio-Mimetic Clam-Like Swimmer in a 3D Incompressible Fluid

In this case we will consider the model of a bio-mimetic swimmer as in Fig. 4.4. Once again, while in Fig. 3.5, the parts of the swimmer's body are identical parallelepipeds, in the general case they are assumed to be as in Assumption 2.1.

Similar to the derivation of models (3.2.1)–(3.2.2) in Sect. 3.2, again under Assumption 2.1, we can obtain the following model for our clam-like artificial swimmer in a fluid described by the 3D non-stationary Stokes equations with *controlled elastic forces* (Hooke's elastic forces can also be considered, of course):

$$\begin{cases} u_t = \nu \Delta u + F_{clam3D} - \nabla p & \text{in } Q_T = (0, T) \times \Omega, \\ \operatorname{div} u = 0 & \text{in } Q_T, \\ u = 0 & \text{in } \Sigma_T = (0, T) \times \partial\Omega, \\ u(0, \cdot) = u_0 & \text{in } \Omega \subset R^3, \\ \frac{dz_i}{dt} = \frac{1}{\operatorname{meas}(S)} \int_{S(z_i(t))} u(t, x)dx, \quad z_i(0) = z_{i,0}, \quad i = 1, 2, 3, \end{cases}$$

$$\tag{4.4.1}$$

z_1 (t)

z_3 (t)

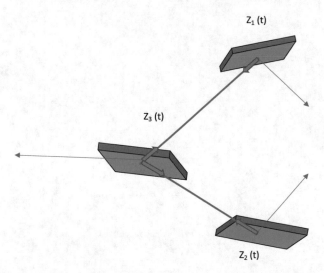

z_2 (t)

Fig. 4.4 A 3D clam-like bio-mimetic swimmer that consists of 3 parallelepipeds with all internal forces engaged

where

$$z(t) = (z_1(t), z_2(t), z_3(t)) \in [R^2]^3, \quad w(t) = (w_1(t), w_2(t)),$$

$$F_{clam3D}(t, x) = v_1(t)\left[\xi_1(t, x)\Lambda_i(t)(z_1(t) - z_3(t))\right.$$

$$\left. -\xi_2(t, x)\frac{\|z_1(t) - z_3(t)\|^2}{\|z_2(t) - z_3(t)\|^2}B_i(t)(z_2(t) - z_3(t))\right]$$

$$+ \xi_3(t, x)v_1(t)\left[A_i(t)(z_3(t) - z_1(t)) - \frac{\|z_1(t) - z_3(t)\|^2}{\|z_2(t) - z_3(t)\|^2}B_i(t)(z_3(t) - z_2(t))\right]$$

$$+ \xi_1(t, x)w_1(z_3(t) - z_1(t)) + \xi_3(t, x)w_1(z_1(t) - z_3(t))$$

$$+ \xi_2(t, x)w_2(z_3(t) - z_2(t)) + \xi_3(t, x)w_2(z_2(t) - z_3(t)). \tag{4.4.2}$$

In the case of the Navier–Stokes equations, we need to replace (4.4.1) with the following hybrid coupled system:

$$\begin{cases} u_t + (u \cdot \nabla)u = \nu \Delta u + F_{clam3D} - \nabla p & \text{in } Q_T = (0, T) \times \Omega, \\ \operatorname{div} u = 0 & \text{in } Q_T, \\ u = 0 & \text{in } \Sigma_T = (0, T) \times \partial\Omega, \\ u(0, \cdot) = u_0 & \text{in } \Omega \subset R^3, \\ \frac{dz_i}{dt} = \frac{1}{\operatorname{meas}(S)}\int_{S(z_i(t))} u(t, x)dx, \quad z_i(0) = z_{i,0}, \quad i = 1, 2, 3. \end{cases}$$

$$\tag{4.4.3}$$

Part II
Well-Posedness of Models for Bio-Mimetic Swimmers in 2D and 3D Incompressible Fluids

Chapter 5
Well-Posedness of 2D or 3D Bio-Mimetic Swimmers: The Case of Stokes Equations

In this chapter we will investigate the mathematical well-posedness of a variety of empiric models introduced for swimmers in Chaps. 2–4 in the case of a fluid governed by either 2D or 3D non-stationary Stokes equations, which are often regarded as a linearized version of Navier–Stokes equations. The content of this chapter is a principal development of works [19, 25, 26, 28, 29, 31], which do not assume geometric controls.

5.1 Notations

Below, we use the following notations:

- The symbol $\Omega \subseteq R^K$, $K = 2, 3$, denotes a bounded spatial domain (open simply connected set) for fluid with continuously twice differentiable boundary $\partial\Omega$. (Such assumption on the boundary is to avoid some technicalities and it can be relaxed in the situations when, e.g., we do not deal with regularity in the space $(H^2(\Omega)^2$, introduced below in this section.)
- $Q_T = (0, T) \times \Omega$.
- The symbol S is reserved for a basic set which is used to construct swimmers' bodies, see Sect. 5.2 for more details.
- $C^\infty(\Omega)$ denotes the space of infinitely many times differentiable functions with compact support in Ω.
- $\mathscr{D}'(\Omega)$ denotes space of distributions in Ω, i.e., the dual space of $C^\infty(\Omega)$.
- $W^{1,p}(\Omega)$, $1 \leq p \leq \infty$, denote the Sobolev spaces over Ω, i.e., the Banach spaces of functions in $L^p(\Omega)$ whose first (generalized) derivatives belong to $L^p(\Omega)$.
- $H^1(\Omega) = W^{1,2}(\Omega)$, and $H^2(\Omega) = W^{2,2}(\Omega)$.
- $H^{2,1}(Q_T) = \{\varphi \mid \varphi \in L^2(0, T; H^2(\Omega)), \ \varphi_t \in L^2(Q_T)\}$.
- $H^1_0(\Omega)$ denotes the subspace of $H^1(\Omega)$ consisting of functions vanishing on $\partial\Omega$.

© The Author(s), under exclusive license to Springer Nature Switzerland AG 2021
A. Khapalov, *Bio-Mimetic Swimmers in Incompressible Fluids*, Lecture Notes
in Mathematical Fluid Mechanics, https://doi.org/10.1007/978-3-030-85285-6_5

- $H^{-1}(\Omega)$ denotes the dual space of $H_0^1(\Omega)$.
- The norms of spaces that are Cartesian products (vectors) of normed spaces are equipped with the standard (Pythagorean) product norm, i.e., if $\phi = (\phi_1, \ldots, \phi_L) \in (X)^L = [X]^L$, where X is a normed space, then $\| \phi \|_{(X)^L} = \sqrt{\sum_{l=1}^{L} \| \phi_l \|_X^2}$ (also having in mind that all norms on finite-dimensional vector spaces are equivalent to one another).
- $\mathcal{V} = \{\varphi \in [C^\infty(\Omega]^K \,|\, \mathrm{div}\,\varphi = 0\}$.
- H is the closure of $\mathcal{V}$ in the $(L^2(\Omega))^K$-norm and V is the closure of $\mathcal{V}$ in $(H_0^1(\Omega))^K$-norm, defined by the following scalar product:

$$[\phi_1, \phi_2] = \int_\Omega \sum_{j=1}^{3} \sum_{i=1}^{3} \phi_{1 j x_i} \phi_{2 j x_i}\, dx,$$

$$\phi_1(x) = (\phi_{11}, \phi_{12}, \phi_{13}), \quad \phi_2(x) = (\phi_{21}, \phi_{22}, \phi_{23}).$$

- $G(\Omega)$ is the orthogonal complement of H in $(L^2(\Omega))^K$.
- Let V' and H' stand for the dual spaces, respectively, of V and H. Then, identifying H with H', we have

$$V \subset H \equiv H' \subset V'.$$

- Below we will use the notations $\|\varphi\|_{L^2}$ and $\|\varphi\|_{H_0^1}$, respectively, both for functions $\varphi \in L^2(\Omega)$ and $\varphi \in H_0^1(\Omega)$ and for functions $\varphi \in (L^2(\Omega))^K$ and $\varphi \in (H_0^1(\Omega))^K$, and similar spaces.
- We use notation $|a|$ for norms of elements of R^K.
- As usual, $B_\varepsilon(x_0)$ will be used to denote an open ball of radius $\varepsilon > 0$ with center at x_0 in a respective metric space.
- Below and above, to simplify notations, we will use both $F(z)$ (or just F) and $F(t, x) = (F(z))(t, x)$ to denote the force term in a fluid equation.

5.2 Swimmer's Body

In addition to Assumption 2.1 and the discussion in Sect. 2.5, we will use below (in all chapters of this monograph) the following two assumptions:

Assumption (A5.1)
1. *Each of the sets $S(z_i(t))$, $i = 1, \ldots, n$, $t \in [0, T]$, forming a body of a swimmer at hand, can be obtained by shifting and rotating (we call the latter—"applying a geometric control") a prescribed (open connected) set*

$$S \subset B_r(0) \subset R^K,$$

centered about the origin, where a prescribed value $r > 0$ is "very small" relative to the size of Ω, namely,

$$S(z_i(t)) = S_i(0, t) + z_i(t), \quad i = 1, \ldots, n, \quad t \in [0, T]. \tag{5.2.1}$$

Each of the sets $S_i(0, t)$'s can be obtained by rotating S about its center ("0" stands for the origin), so that the sets $\{(t, x) | x \in S(z_i(t)), t \in (0, T)\} \subset R^{1+K}, i = 1, \ldots, n,$ are measurable for any $z_i \in C([0, T])^K$, see also Remark 5.1 below for more details.

2. *If Hooke's elastic forces (2.3.3) are available for a swimmer, we also assume that*

$$2r < l_i, \quad i = 1, \ldots, n - 1.$$

Remark 5.1 (Geometric Controls)

- Thus, each geometric control is an *individually prescribed* time-dependent rotation mapping of respective $S(z_i(t))$ in time (Figs. 5.1 and 5.2):

$$t \longrightarrow S_i(0, t) = S(z_i(t)) - z_i(t) \subset R^K, \quad t \in [0, T],$$

$$i = 1, \ldots, n, \quad \forall z_i \in C([0, T]; R^K). \tag{5.2.2}$$

- If all the sets $S(z_i(t))$'s do not change their initial spatial orientation on $[0, T]$, we will say that the swimmer does not engage its geometric controls.
- As we mentioned earlier in Remark 1.1 in Chap. 1, we assume that the effect of acting geometric controls (i.e., rotations of $S(z_i(t))$'s) themselves on the fluid velocity, and, hence, on the motion of swimmers in this fluid, is negligible. In turn, using them in combination with the internal control forces is instrumental for the swimmers' locomotion, see Chaps. 9–10.

Given a unit vector $\eta \in R^K, |\eta| = 1$, for every $h \in R^K$ and $y \in Y_\eta \subset R^K$, where Y_η is a hyperplane perpendicular to η,

$$Y_\eta = \{y \mid y \cdot \eta = 0\},$$

introduce the following two one-dimensional subsets in Ω:

$$(S)_\eta^y = \{y + t\eta \mid y + t\eta \in \Omega, t \in R\}, \, (S_h)_\eta^y = (S)_\eta^y \bigcap S_h, \tag{5.2.3}$$

where $S_h = (h + S(0)) \, \Delta \, S(0)$ is the symmetric difference between $S(0)$ and $h + S(0)$, i.e.,

$$S_h = ((h + S) \cup S) \setminus ((h + S) \cap S).$$

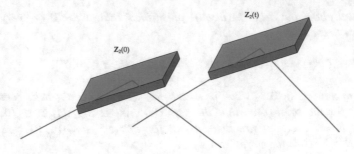

Fig. 5.1 A 3D swimmer that does *not engage a geometric control* for $S(z_2(t))$: the point $z_2(t)$ moves in time, while the spatial orientation of $S(z_2(t))$ about $z_2(t))$ remains the same

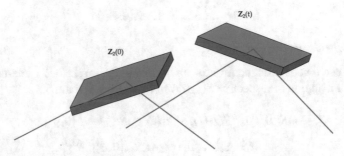

Fig. 5.2 A 3D swimmer that *engages a geometric control* for $S(z_2(t))$: the point $z_2(t)$ moves in time, and the spatial orientation of $S(z_2(t))$ about $z_2(t))$ changes in time as well

In other words, every set $(S_h)^y_\eta$ is a one-dimensional set on the line $y + t\eta, t \in R$, drawn parallel to η through the point $y \in Y_\eta$, that lies within S_h.

Assumption (A5.2) (Lipschitz Continuity of Shifts of Swimmer's Body) *There exist positive constants h_0 and $\mathcal{M}$ such that for any vector $h \in B_{h_0}(0)$, we can find a vector $\eta = \eta(h), \mid \eta \mid = 1$, that satisfies the following uniform estimate:*

$$\text{meas } \{(S_h)^y_\eta\} \; = \; \int_{(S_h)^y_\eta} dx \; \leq \; \mathcal{M} \, |h| \qquad \forall \, y \in Y_\eta. \tag{5.2.4}$$

Note that estimate (5.2.4) applies to sets $S(z_i(t))$ from Assumption (A5.1) at any time t in place of S_h.

In the case when no geometric controls are engaged (i.e., no changes in geometric orientation of $S(z_i(t))$, $t \in [0, T]$, $i = 1, \ldots, n$) and η is parallel to h, **(A5.2)** holds, e.g., for discs and rectangles in 2D and for balls and parallelepipeds in 3D.

Remark 5.2

- Assumption (A5.2) is weaker than just to assume that $\eta = h$, see, e.g., an illustrative example in [33] for a "saw"-like set S.

- Under Assumption 2.1, assumed everywhere in this monograph, the sets $(S)_\eta^y$ are open uniformly bounded segments (they can be empty).

Integrity of Swimmers' Bodies   Our results below will be established to ensure the following natural restrictions on the integrity of swimmers's bodies (see also Chap. 2):

- All swimmers lie in Ω at all times with an r-margin:

$$\bigcup_{i=1}^{n} B_{2r}(z_i(t)) \subset B_{\mathscr{S}}(a(t)) \subset \Omega \quad \forall t, \ \ i = 1, \ldots, n, \tag{5.2.5}$$

where $B_{\mathscr{S}}(a(t))$ is an open ball in R^K, whose center's position $a(t)$ can change in time, but the radius $\mathscr{S}$ is fixed for each swimmer and serves as a restriction for its maximum allowed size;
- The body parts of swimmers cannot overlap:

$$|z_i(t) - z_j(t)| > 2r \quad \forall t, \ \ i \neq j, \ \ i, j = 1, \ldots, n. \tag{5.2.6}$$

On How to Ensure (5.2.5)–(5.2.6): Strategy of Proofs Below   In our proofs below, we will follow the following strategy:

1. We will study first the uncoupled PDE and ODE equations in our model (5.3.1) below, with suitable initial states for which (5.2.5)–(5.2.6) hold.
2. Then, we will establish the existence of their solutions on some $(0, T_1)$ such that (5.2.5)–(5.2.6) will hold on $(0, T_1)$.
3. Next, we will use a fixed point theorem to prove the existence of a "fixed point" solution for the coupled system (5.3.1), among the aforementioned decoupled solutions.
4. After that, we will investigate how we can extend such a solution form $(0, T_1)$ to a desirable time interval $(0, T)$.
5. To this end, we will also use the rotational and controlled elastic swimmer's forces, along with its geometric controls, to ensure that the body of the swimmer at hand preserved its desirable dimensions and configuration. *Thus, the question of global existence of solutions is linked, in this monograph, to the issue of suitable selection of coefficients v_i's and w_i's and geometric controls.*

5.3   Initial- and Boundary-Value Problem Setup

In this chapter we will consider the following initial- and boundary-value problem suitable to describe all the models introduced in Part I of this monograph:

$$
\begin{cases}
u_t = \nu \Delta u + F - \nabla p & \text{in } Q_T = (0, T) \times \Omega, \\
\operatorname{div} u = 0 & \text{in } Q_T, \\
u = 0 & \text{in } \Sigma_T = (0, T) \times \partial\Omega, \\
u(0, \cdot) = u_0 & \text{in } \Omega \subset R^K, \quad K = 2, 3, \\
\frac{dz_i}{dt} = \frac{1}{\operatorname{meas}(S)} \int_{S(z_i(t))} u(t, x) dx, & z_i(0) = z_{i,0}, \;\; i = 1, \ldots, n,
\end{cases}
$$

$$(5.3.1)$$

where $x = (x_1, \ldots, x_K)$, $u(t, x) = (u_1(t, x), \ldots, u_K(t, x))$,

$$
z(t) = (z_1(t), \ldots, z_n(t)) \in [R^K]^n, \quad v(t) = (v_1(t), \ldots, v_{n-2}(t)) \in R^{n-2},
$$

$$
w(t) = (w_1(t), \ldots, w_{n-1}(t)) \in R^{n-1},
$$

a set of geometric controls

$$
S_i(0, t), \quad t \in [0, T], \quad i = 1, \ldots, n, \tag{5.3.2}
$$

is prescribed, and F can be of any configuration of the swimmers' internal rotational and elastic forces introduced in Chaps. 2–4 in 2D and 3D cases (as marked by respective subscripts below), namely,

$$
F_{rot2D}(t, x) = \sum_{i=2}^{n-1} v_{i-1}(t) \Bigg[\xi_{i-1}(t, x) A(z_{i-1}(t) - z_i(t))
$$

$$
- \xi_{i+1}(t, x) \frac{\|z_{i-1}(t) - z_i(t)\|^2}{\|z_{i+1}(t) - z_i(t)\|^2} A(z_{i+1}(t) - z_i(t)) \Bigg]
$$

$$
+ \sum_{i=2}^{n-1} \xi_i(t, x) v_{i-1}(t) \Bigg[A(z_i(t) - z_{i-1}(t)) - \frac{\|z_{i-1}(t) - z_i(t)\|^2}{\|z_{i+1}(t) - z_i(t)\|^2} A(z_i(t) - z_{i+1}(t)) \Bigg],
$$

$$(5.3.3)$$

$$
F_{cef2Dor3D}(t, x) = \sum_{i=2}^{n} [\xi_{i-1}(t, x) w_{i-1}(z_i(t) - z_{i-1}(t)) + \xi_i(t, x) w_{i-1}(z_{i-1}(t) - z_i(t))],
$$

$$(5.3.4)$$

$$
F_{hef2Dor3D}(t, x) = \sum_{i=2}^{n} [\xi_{i-1}(t, x) k_{i-1} \frac{(\|z_i(t) - z_{i-1}(t)\|_{R^2} - l_{i-1})}{\|z_i(t) - z_{i-1}(t)\|_{R^2}} (z_i(t) - z_{i-1}(t))
$$

$$
+ \xi_i(t, x) k_{i-1} \frac{(\|z_i(t) - z_{i-1}(t)\|_{R^2} - l_{i-1})}{\|z_i(t) - z_{i-1}(t)\|_{R^2}} (z_{i-1}(t) - z_i(t))], \tag{5.3.5}
$$

$$
F_{rot3D}(t, x) = \sum_{i=2}^{N-1} v_{i-1}(t) \Bigg[\xi_{i-1}(t, x) \; A_i(t) (z_{i-1}(t) - z_i(t))
$$

$$-\xi_{i+1}(t,x)\,\frac{|z_{i-1}(t)-z_i(t)|^2}{|z_{i+1}(t)-z_i(t)|^2}\,B_i(t)\big(z_{i+1}(t)-z_i(t)\big)\Bigg]$$

$$+\sum_{i=2}^{N-1}\xi_i(t,x)\,v_{i-1}(t)\Bigg[A_i(t)\big(z_i(t)-z_{i-1}(t)\big)-\frac{|z_{i-1}(t)-z_i(t)|^2}{|z_{i+1}(t)-z_i(t)|^2}\,B_i(t)\big(z_i(t)-z_{i+1}(t)\big)\Bigg].$$

$$(5.3.6)$$

5.3.1 Estimates for Internal Forces

Making use of the formula for the cross product:

$$\|a\times b\|_{R^K}=\|a\|_{R^K}\|b\|_{R^K}\,|\sin\Theta|,$$

where Θ is an angle between vectors a and b in R^K, we can derive the following set of estimates.

Lemma 5.1 *Assume that the above restrictions (5.2.5)–(5.2.6) hold and that $z\in[C([0,T];R^K)]^n$, $K=2,3$, $v=(v_1,\ldots,v_{n-2})\in[L^\infty(0,T)]^{n-2}$, and $w=(w_1,\ldots,v_{n-1})\in[L^\infty(0,T)]^{n-2}$. Then, the parts of the forcing term $F(t,x)$ (in any configuration composed form (5.3.3)–(5.3.6)) in (5.3.1) lie in $L^\infty(0,T;[L^2(\Omega)]^K)$, while the following estimates hold:*

$$\|F_{hef2Dor3D}\|_{[L^\infty(Q_T)]^K}\le C\max_{i=2,\ldots,n}\Big\{\|z_i-z_{i-1}\|_{C([0,T];R^K)}+l_{i-1}\Big\}\sum_{i=2}^{n}|k_{i-1}|,$$

$$\|F_{cef2Dor3D}\|_{[L^\infty(Q_T)]^K}\le C\max_{i=2,\ldots,n}\|z_i-z_{i-1}\|_{C([0,T];R^K)}\sum_{i=2}^{n}\|w_{i-1}\|_{L^\infty(0,T)},$$

$$\|F_{rot2D}\|_{[L^\infty(Q_T)]^K}\le C\sum_{i=2}^{n-1}\|v_{i-1}\|_{L^\infty(0,T)}$$

$$\times\max_{i=2,\ldots,n-1}\left\{\|z_i-z_{i-1}\|_{C([0,T];R^2)}+\frac{\|z_i-z_{i-1}\|^2_{C([0,T];R^2)}}{2r}\right\},$$

$$\|F_{rot3D}\|_{[L^\infty(Q_T)]^K}\le C\sum_{i=2}^{n-1}\|v_{i-1}\|_{L^\infty(0,T)}\max_{i=2,\ldots,n}\|z_i-z_{i-1}\|^3_{C([0,T];R^3)},$$

$$(5.3.7)$$

where C > 0 is a (generically denoted) constant. These estimates yield that

$$\|F\|_{[L^\infty(Q_T)]^K} \leq \zeta, \tag{5.3.8}$$

where a positive constant $\zeta > 0$ depends on the size of Ω in R^K and the $L^\infty(0, T)$-norms of v_i's and w_i's and on k_i's.

Remark 5.3

- In this monograph, we select $L^\infty(0, T)$ as the main function space for $v_i(t)$'s and $w_i(t)$'s as they are to define the magnitudes of *bounded* swimmer's internal forces in empiric models introduced in Chaps. 2–4. However, from an abstract mathematical viewpoint, other function spaces (e.g., $L^2(0, T)$) can also be considered.
- In this monograph, we regard $v_i(t)$'s and $w_i(t)$'s as *selectable control parameters or controls.*

5.4   Main Result: Existence and Uniqueness of Solutions

Here are the main well-posedness results of this chapter, where we establish the existence and uniqueness of solutions to system (5.3.1) on some interval $(0, T^*)$, assuming that, during this time interval, *a swimmer can engage prescribed geometric controls.* (For the case, when a swimmer does not use geometric controls, we refer to [19, 25, 26, 28, 29, 31], where various well-posedness results were derived for 2D and 3D incompressible fluids, governed by the non-stationary Stokes equations.)

Theorem 5.1 *Let $u_0 \in V$; $T > 0$; a set of geometric controls $S_i(0, t)$, $t \in [0, T]$, $i = 1, \ldots, n$; $k_i > 0$, $i = 1, \ldots, n - 1$; $v \in [L^\infty(0, T)]^{n-2}$; $w \in [L^\infty(0, T)]^{n-1}$ and $z_i(0) \in \Omega \subset R^K$, $i = 1, \ldots, n$, , $K = 2, 3$ be given, Assumptions (A5.1)-(A5.2) hold and structural restrictions (5.2.5)–(5.2.6) hold as well at $t = 0$. Let us assume that some geometric controls are prescribed for each of the sets $S(z_i(t)), t \in [0, T]$. Then there exists a $T^* \in (0, T]$ such that the system (5.3.1) admits a unique solution $\{u, p, z\}$ on $(0, T^*)$, $\{u, \nabla p, z\} \in L^2(0, T^*; H) \times L^2(0, T^*; G(\Omega)) \times [C([0, T^*]; R^K)]^n$, $K = 2, 3$, $u \in C([0, T^*]; V)$. The restrictions (5.2.5)–(5.2.6) hold on $[0, T_*]$.*

It will follow from the proof below that solutions to (5.3.1) can be extended in time as long as the restrictions (5.2.5)–(5.2.6) continue to hold.

The regularity results from [36, 37] yield the following refinement of Theorem 5.1 under our assumption that $\partial\Omega$ is twice continuously differentiable (see Sect. 5.1).

Corollary 5.1 *Under the conditions of Theorem 5.1, $u_t \in (L^2(Q_{T^*}))^K$ and $u_{x_i x_j}, p_{x_i} \in (L^2(Q_{T^*}))^K$.*

Corollary 5.2 *Making use of estimate similar to (6.2.7), in Theorem 5.1, we can also consider $u_0 \in H$, in which case in Theorem 5.1 we will have $u \in C(0, T^*; H)$ (compare to Theorem 6.1 below).*

5.5 Proof of Theorem 5.1

Below, to be explicit in notations, we focus on the 3D case. The 2D case is similar. Our plan of proof is as follows:

- In Sect. 5.5.1, we discuss the existence and uniqueness of the solutions to the decoupled version of the equation for $z_i(t)$'s in (5.3.1).
- In Sect. 5.5.2, we will introduce three continuous mappings for the decoupled version of the system (5.3.1) and will prove that one of them is compact.
- In Sect. 5.5.3, we will apply a fixed point argument to prove Theorem 5.1.

Remark 5.4 To simplify the notations,

1. throughout this monograph, we use *the symbol "C" as a generic notation to denote possibly different constants*;
2. throughout the proof below, we assume that all assumptions of Theorem 5.1 hold.

5.5.1 Preliminary Results: Decoupled Equation for $z_i(t)$'s

Introduce the following decoupled version of equations for $z_i(t)$'s in (5.3.1):

$$\frac{dp_i}{dt} = \frac{1}{\text{meas}(S)} \int_{S(p_i(t))} u(t, x) dx, \quad p_i(0) = z_{i,0}, \quad i = 1, \ldots, n, \quad t \in (0, T),$$

$$(5.5.1)$$

where $u(t, x)$ and geometric controls are given for each $S(p_i(t)), i = 1, \ldots, n, t \in [0, T]$, and let $p(t) = (p_1(t), \ldots, p_n(t))$.

Lemma 5.2 *Let $T > 0$ and $u \neq 0, u \in L^2(0, T; V)$ be given. Then there is a $T_0 \in (0, T]$ such that system (5.5.1) admits a unique solution in $C([0, T_0]; R^3)$ satisfying conditions (5.2.5)–(5.2.6) with $p(t)$ in place of $z(t)$, if they hold at time $t = 0$ for $z_{i,0}, i = 1, \ldots, n$.*

Proof of Existence Let $h_0 > 0$ be such that the restrictions (5.2.5)–(5.2.6) would hold for any $a_{i,0} \in B_{h_0/2}(z_{i,0}) \subset R^3, i = 1, \ldots, n$, placed in (5.2.5)–(5.2.6) instead of $z_{i,0}, i = 1, \ldots, n$.

Select any T_0 such that

$$0 < T_0 < \min\left\{\frac{\text{meas}(S)h_0^2}{4\|u\|_{(L^2(Q_T))^3}^2} T, 1\right\},$$

$$(5.5.2)$$

and introduce the following auxiliary mapping:

$$D_i : C([0, T_0]; R^3) \supset \mathscr{B}_{h_0/2}(z_{i0}) \longrightarrow C([0, T_0]; R^3),$$

$$D_i(p_i(t)) = z_{i0} + \frac{1}{\text{meas}(S)} \int_0^t \int_{S(p_i(\tau))} u(\tau, x) dx d\tau, \quad i = 1, \dots, n, \ t \in [0, T_0].$$

Then, in view of (5.5.2),

$$\|D_i(p_i(t)) - z_{i0}\|_{C([0,T_0]; R^3)} \leq \frac{\sqrt{T_0}}{\sqrt{\text{meas}(S)}} \|u\|_{(L^2(Q_T))^3} < \frac{h_0}{2}. \tag{5.5.3}$$

Thus, D_i *maps* $\mathscr{B}_{h_0/2}(z_{i0})$ *into itself for each* $i = 1, \dots, n$.

Continuity of Operators D_i's Consider any two functions $p_i^{(1)}(t)$ and $p_i^{(2)}(t) \in \mathscr{B}_{h_0/2}(z_{i0})$.

Let $\xi(x, \mathscr{S})$ denote the characteristic function of a set $\mathscr{S} \subset R^3$. Then, making use of Assumption (A5.2), we obtain for $t \in [0, T_0]$

$$\|D_i(p_i^{(1)}(t)) - D_i(p_i^{(2)}(t))\|_{R^3}^2$$

$$= \frac{1}{\text{meas}^2(S)} \left\| \int_0^t \int_{S(p_i^{(1)}(\tau))} u(\tau, x) dx d\tau - \int_0^t \int_{S(p_i^{(2)}(\tau))} u(\tau, x) dx d\tau \right\|_{R^3}^2$$

$$= \frac{1}{\text{meas}^2(S)} \left\| \int_0^t \int_{\Omega} u(\tau, x) \left(\xi(x, S_i(p_i^{(1)}(\tau))) - \xi(x, S(p_i^{(2)}(\tau))) \right) dx d\tau \right\|_{R^3}^2$$

$$\leq \frac{1}{\text{meas}^2(S)} \sum_{j=1}^3 \left[\int_0^t \int_{\Omega} | u_j(\tau, x) | \, | \xi(x, S(p_i^{(1)}(\tau))) - \xi(x, S(p_i^{(2)}(\tau))) | \, dx d\tau \right]^2, \tag{5.5.4}$$

where $u = (u_1, u_2, u_3)$.

Now, for every $\tau \in (0, t)$, we can split each integral over Ω in (5.5.4) into a three one-dimensional integrals (along Fubini's theorem), one of which can be selected to be along the vector $\eta = \eta(\tau)$ (as defined by the prescribed geometric control (5.2.2)) in Assumption (5.2) for the shift of the set $S(p_i^{(1)}(\tau))$ along the vector

$$h = p_i^{(2)}(\tau)) - p_i^{(1)}(\tau)). \tag{5.5.5}$$

Then,

1. we can apply estimate (5.2.4) in Assumption (A5.2) along with
2. the continuous embedding theorem $C[a_*, b_*]$ into $H_0^1(a_*, b_*)$, applied to evaluate $| u_j(\tau, x) |$ in (5.5.4) along segments $(a_*(\tau, y), b_*(\tau, y))$, $(S_h)_\eta^y \subset$

$(a_*(\tau, y), b_*(\tau, y)) \subset (S)^y_\eta$ such that $a_*(\tau, y), b_*(\tau, y) \in \partial\Omega$ (note that the length of $(a_*(\tau, y), b_*(\tau, y))$'s is bounded by the diameter of Ω),

to obtain (see (5.5.7) for details)

$$\|D_i(p_i^{(1)}(t)) - D_i(p_i^{(2)}(t))\|^2_{R^3}$$

$$\leq \frac{C^2 t}{\text{meas}^2(S)} \|u\|^2_{L^2(0,T;V)} \|p_i^{(1)} - p_i^{(2)}\|^2_{C([0,T_0];R^3)}, \quad t \in [0, T_0], \qquad (5.5.6)$$

where C^2 depends on the diameter of Ω and $\mathcal{M}$ in (5.2.4) (see also Remark 5.4), which implies the required continuity of D_i's.

In the above we used estimates of the following type:

$$\frac{1}{\text{meas}^2(S)} \left[\int_0^t \int_\Omega | u_j(\tau, x) | \, | \xi(x, S(p_i^{(1)}(\tau))) - \xi(x, S(p_i^{(2)}(\tau))) | \, dxd\tau \right]^2$$

$$= \frac{1}{\text{meas}^2(S)} \left[\int_0^t \int\int \int_{(S_h)^y_\eta} |u_j(\tau, x)| \, | \xi(x, S(p_i^{(1)}(\tau))) - \xi(x, S(p_i^{(2)}(\tau))) | \, dxd\tau \right]^2$$

$$\leq \frac{1}{\text{meas}^2(S)} \left[\int_0^t \int\int C(y) \|u_j(\tau, \cdot)\|_{H^1_0(a_*(\tau,y), b_*(\tau,y))} \right.$$

$$\left. \times \int_{(S_h)^y_\eta} | \xi(x, S(p_i^{(1)}(\tau))) - \xi(x, S(p_i^{(2)}(\tau))) | \, dxd\tau \right]^2$$

$$\leq \frac{\mathcal{M}^2}{\text{meas}^2(S)} \left[\int_0^t \int\int C(y) \|u_j(\tau, \cdot)\|_{H^1_0(a_*(\tau,y), b_*(\tau,y))} \|p_i^{(1)} - p_i^{(2)}\|_{C([0,T_0];R^3)} dyd\tau \right]^2$$

$$\leq \frac{C^2}{\text{meas}^2(S)} \left[\int_0^t \|u_j(\tau, \cdot)\|_{H^1_0(\Omega)} \|p_i^{(1)} - p_i^{(2)}\|_{C([0,T_0];R^3)} dxd\tau \right]^2, \quad t \in [0, T_0].$$
$$(5.5.7)$$

Equicontinuity of Sets $D_i(p_i)(\mathcal{B}_{h_0/2}(z_{i0}))$'s Indeed, consider any $t, t + \delta \in [0, T_0]$, e.g., when $\delta > 0$ (the case when $\delta < 0$ is similar). Then, for $i = 1, \ldots, n$,

$$\|D_i(p_i(t + \delta)) - D_i(p_i(t))\|_{R^3} = \frac{1}{\text{meas}(S)} \left\| \int_t^{t+\delta} \int_{S(p_i(\tau))} u(\tau, x)dxd\tau \right\|_{R^3}$$

$$\leq \frac{\sqrt{\delta}}{\sqrt{\text{meas}(S)}} \|u\|_{(L^2(Q_T))^3}, \quad j = 1, \ldots, \tag{5.5.8}$$

which implies the desirable equicontinuity. Therefore, by the Arzela–Ascoli theorem, the sets $D_i(p_i)(\mathscr{B}_{h_0/2}(z_{i0}))$'s are compact, and, thus, the operators D_i's are compact mappings from $\mathscr{B}_{h_0/2}(z_{i0}) \Rightarrow \mathscr{B}_{h_0/2}(z_{i0})$.

Applying Schauder's fixed point theorem yields that there exist $p_i(t) \in C([0, T_0]; R^3)$, $i = 1, \ldots, n$, such that $D_i(p_i(t)) = p_i(t)$, i.e., (5.5.1) holds.

Uniqueness of Solutions to (5.5.1) It follows form (5.5.6).

Restrictions (5.2.5)–(5.2.6) Hold Restrictions (5.2.5)–(5.2.6) hold due to estimate (5.5.3) due to our selection of h_0.

This ends the proof of Lemma 5.2.

5.5.2 Three Decoupled Solution Mappings for (5.3.1)

Let $\mathscr{B}_q(0)$ denote a closed ball of radius q (its value will be selected later) with center at the origin in the space $\mathscr{H} = L^2(0, T; V) \bigcap L^2(0, T; H)$ endowed with the following norm:

$$\|\phi\|_{\mathscr{H}} = \sqrt{\|\phi\|^2_{L^2(0,T;(L^2(\Omega))^3)} + \|\phi\|^2_{L^2(0,T;V)}},$$

thus,

$$\mathscr{B}_q(0) = \{\phi \in \mathscr{H} \mid \|\phi\|_{\mathscr{H}} \leq q\}.$$

Note that Lemma 5.2 holds for any $u \in \mathscr{B}_q(0)$.

5.5.2.1 Solution Mapping A for $z_i(t)$, $i = 1, \ldots, n$

We now intend to show that the operator

$$\mathbf{A} : \mathscr{B}_q(0) \longrightarrow [C([0, T]; R^3)]^n, \quad \mathbf{A}u = p = (p_1, \ldots, p_n),$$

where the p_i's solve (5.5.1), is continuous and compact if $T > 0$ is sufficiently small.

Continuity of A Let $u^{(1)}, u^{(2)} \in \mathscr{B}_q(0)$ with $T = T_0$ from Lemma 5.2.

Define $\mathbf{A}u^{(j)} = w^{(j)} = (w_1^{(j)}, \ldots, w_n^{(j)})$ for $j = 1, 2$. To show that $\mathbf{A}$ is continuous, we will evaluate

$$\|\mathbf{A}u^{(1)} - \mathbf{A}u^{(2)}\|_{[C([0,T_0];R^3)]^n}. \tag{5.5.9}$$

To this end, similar to (5.5.3) and (5.5.6), we have the following estimates for $t \in [0, T_0]$:

$$\|w_i^{(1)}(t) - w_i^{(2)}(t)\|_{R^3}$$

$$= \left\| \frac{1}{\operatorname{meas}(S)} \int_0^t \int_{S(w_i^{(1)}(\tau))} u^{(1)}(x,\tau)dxd\tau - \frac{1}{\operatorname{meas}(S)} \int_0^t \int_{S(w_i^{(2)}(\tau))} u^{(2)}(x,\tau)dxd\tau \right\|_{R^3}$$

$$= \frac{1}{\operatorname{meas}(S)} \left\| \int_0^t \int_{S(w_i^{(1)}(\tau))} u^{(1)}(x,\tau)dxd\tau - \int_0^t \int_{S(w_i^{(1)}(\tau))} u^{(2)}(x,\tau)dxd\tau \right.$$

$$\left. + \int_0^t \int_{S(w_i^{(1)}(\tau))} u^{(2)}(x,\tau)dxd\tau - \int_0^t \int_{S(w_i^{(2)}(\tau))} u^{(2)}(x,\tau)dxd\tau \right\|_{R^3}$$

$$\leq \frac{1}{\operatorname{meas}(S)} \left(\left\| \int_0^t \int_\Omega (u^{(1)}(x,\tau) - u^{(2)}(x,\tau))\xi(x, S(w_i^{(1)}(\tau)))dxd\tau \right\|_{R^3} \right.$$

$$\left. + \left\| \int_0^t \int_\Omega u^{(2)}(x,\tau)(\xi(x, S(w_i^{(1)}(\tau))) - \xi(x, S(w_i^{(2)}(\tau))))dxd\tau \right\|_{R^3} \right)$$

$$\leq \frac{\sqrt{T_0}}{\sqrt{\operatorname{meas}(S)}} \|u^{(1)} - u^{(2)}\|_{(L^2(Q_{T_0}))^3} + \frac{C\sqrt{T_0}}{\operatorname{meas}(S)} \|u^{(2)}\|_{L^2(0,T_0;V)} \|w_i^{(1)}$$

$$- w_i^{(2)}\|_{C([0,T_0];R^3)}. \tag{5.5.10}$$

Select a $T > 0$ as follows:

$$0 < T < \min\left\{ \left(\frac{\operatorname{meas}(S)}{Cq}\right)^2, T_0 \right\}. \tag{5.5.11}$$

Then, replacing T_0 in (5.5.10) with T satisfying (5.5.11) and maximizing the left-hand side of (5.5.10) over $[0, T]$, we obtain

$$\|w_i^{(1)} - w_i^{(2)}\|_{C([0,T];R^3)} \leq \frac{\sqrt{T}}{\sqrt{\operatorname{meas}(S)}} \|u^{(1)} - u^{(2)}\|_{(L^2(Q_T))^3}$$

$$+ \frac{Cq\sqrt{T}}{\operatorname{meas}(S)} \|w_i^{(1)} - w_i^{(2)}\|_{C([0,T_0];R^3)}.$$

In view of (5.5.11), if $w_i^{(1)}(t) \neq w_i^{(2)}(t)$ on $[0, T]$, then the above implies

$$0 < \left(1 - \frac{Cq\sqrt{T}}{\text{meas}(S)}\right) \|w_i^{(1)} - w_i^{(2)}\|_{C([0,T];R^3)} \leq \frac{\sqrt{T}}{\sqrt{\text{meas}(S)}} \|u^{(1)} - u^{(2)}\|_{(L^2(Q_T))^3}.$$

Thus, it follows that

$$\|w_i^{(1)} - w_i^{(2)}\|_{C([0,T];R^3)} \leq \frac{\sqrt{T\,\text{meas}(S)}}{\text{meas}(S) - Cq\sqrt{T}} \|u^{(1)} - u^{(2)}\|_{(L^2(Q_T))^3}. \qquad (5.5.12)$$

Therefore, (5.5.11) and (5.5.12) imply that for every $u^{(1)}, u^{(2)} \in \mathscr{B}_q(0)$,

$$\|\mathbf{A}u^{(1)} - \mathbf{A}u^{(2)}\|_{[C([0,T];R^3)]^n} \leq \frac{\sqrt{nT\,\text{meas}(S)}}{\text{meas}(S) - Cq\sqrt{T}} \|u^{(1)} - u^{(2)}\|_{(L^2(Q_T))^3}.$$

Thus, the operator $\mathbf{A}$ is continuous on $\mathscr{B}_q(0)$ for sufficiently small T as in (5.5.11).

Compactness of A To show that $\mathbf{A}$ is compact, we will show that $\mathbf{A}$ maps any sequence in $\mathscr{B}_q(0)$ into a sequence in $[C([0,T]; R^3)]^n$, which contains a convergent subsequence.

Consider any sequence $\{u^{(j)}\}_{j=1}^{\infty}$ in $\mathscr{B}_q(0)$ with T as in (5.5.11). Let

$$\mathbf{A}u^{(j)} = w_i^{(j)} \quad i = 1, \ldots, n.$$

Let us now show that the functions $w_i^{(j)}, j = 1, \ldots, \infty$, form a family of uniformly bounded and equicontinuous functions. Indeed, applying (5.5.9) to an estimate like (5.5.3) and then maximizing over $[0, T]$ yield

$$\|w_i^{(j)}\|_{C([0,T];R^3)} \leq \max_{i=1,\ldots,n} \left\{\|z_{i0}\|_{R^3}\right\} + \frac{q\sqrt{T}}{\sqrt{\text{meas}(S)}}, \qquad (5.5.13)$$

which provides the aforementioned uniform boundedness.

Consider now any $t, t + h \in [0, T]$, e.g., when $h > 0$ (the case $h < 0$ is similar). Then, for $i = 1, \ldots, n$,

$$\|w_i^{(j)}(t+h) - w_i^{(j)}(t)\|_{R^3} = \frac{1}{\text{meas}(S)} \left\|\int_t^{t+h} \int_{S(w_i^{(j)}(\tau))} u^{(j)}(x, \tau)\,dx\,d\tau\right\|_{R^3}$$

$$\leq \frac{\sqrt{h}}{\sqrt{\text{meas}(S)}} \|u^{(j)}\|_{(L^2(Q_T))^3} \leq \frac{q\sqrt{h}}{\sqrt{\text{meas}(S)}} \quad j = 1, \ldots,$$

which implies the equicontinuity of $\{w_i^{(j)}\}_{j=1}^{\infty}, i = 1, \ldots, n.$

Therefore, by Arzela–Ascoli's theorem, $\{\mathbf{A}u^{(j)}\}_{j=1}^{\infty}$ contains a convergent subsequence in $[C([0, T]; R^3)]^n$, i.e., $\mathbf{A}$ is compact on $\mathscr{B}_q(0)$.

5.5.2.2 Solution Mapping for Decoupled Non-stationary Stokes Equations

Let, for certainty, $T > 0$ be as in (5.5.11). Consider the following decoupled initial- and boundary-value problem for the non-stationary Stokes equations:

$$\frac{\partial y}{\partial t} - \nu \Delta y + \nabla p = f(x, t) \quad \text{in} \ \ Q_T,$$

$$\text{div } y = 0 \ \text{ in } Q_T, \quad y = 0 \ \text{ in } \Sigma_T, \quad y|_{t=0} = y_0 \in V. \tag{5.5.14}$$

For any $f \in (L^2(Q_T))^3$, it is known that the boundary-value problem (5.5.14) admits a unique solution y in $L^2(0, T; V) \cap L^2(0, T; H)$ with the properties described in Theorem 5.1 (see, [36, 37]). Moreover (this can be shown by making use of the respective Fourier series representation, see (7.2.13)), there is a positive constant L such that

$$\|y\|_{L^2(0,T;V)}^2 \le L\|y_0\|_V^2 + L \int_{Q_T} \|f(\cdot, \tau)\|_{R^3}^2 d\tau. \tag{5.5.15}$$

Thus, given y_0, the operator

$$\mathbf{B}f = y, \quad \mathbf{B} : (L^2(Q_T))^3 \longrightarrow L^2(0, T; V) \cap L^2(0, T; H)$$

is continuous.

5.5.2.3 The Force Term

Let $T > 0$ be as in (5.5.11). Given $u \in \mathscr{B}_q(0)$, consider the operator

$$\mathbf{F} : [C([0, T]; R^3)]^n \longrightarrow (L^2(Q_T))^3, \quad \mathbf{F}p = F(p), \ \ p = (p_1, \dots, p_n),$$

where $F(p)$ is from (5.3.1), calculated as in (5.3.3)–(5.3.6).

The aforementioned formulas are explicit elementary algebraic expressions in terms of $(p_i(t) - p_{i-1}(t))$'s. Therefore, after some standard algebraic transformation we can derive along with Lemmas 5.1 and 5.2 (particularly, see (5.5.3)), under the assumptions of Theorem 5.1, that

$$\|F(w^{(1)}) - F(w^{(2)})\|_{(L^2(Q_T))^3} \le M\sqrt{T \text{meas}(\Omega)}\|w^{(1)} - w^{(2)}\|_{[C([0,T];R^3)]^n}, \tag{5.5.16}$$

where M depends, in an increasing on the values of k_i's, l_i's, , the $L^\infty(0, T)$-norms of w and v in the forcing term F, and on

$$\|z_{i0}\|_{R^3} + \frac{\sqrt{T}}{\sqrt{\text{meas}(S)}}q, \quad i = 1, \ldots, n.$$

Hence, $\mathbf{F}$ is a continuous operator for T satisfying (5.5.11).

5.5.3   Proof of Theorem 5.1

Let us summarize the results of Sect. 5.5.2.

We established that for sufficiently small $T > 0$, namely, satisfying (5.5.11), operators

$$\mathbf{A} : \mathscr{B}_q(0) \longrightarrow [C([0, T]; R^3)]^n, \quad \mathbf{A}u = w = (w_1, \ldots, w_n),$$

$$\mathbf{F} : [C([0, T]; R^3)]^n \longrightarrow (L^2(Q_T))^3, \quad \mathbf{F}p = F(p),$$

and

$$\mathbf{B} : (L^2(Q_T))^3 \longrightarrow L^3(0, T; V) \bigcap L^2(0, T; H), \quad \mathbf{B}f = y$$

are all continuous, while $\mathbf{A}$ is also compact. Hence, the operator

$$\mathbf{BFA} : B_q(0) \longrightarrow L^2(0, T; V) \bigcap L^2(0, T; H), \quad \mathbf{BFA}u = y$$

is continuous and compact.

5.5.3.1   Proof of Existence: A Fixed Point Argument

Select the value of q to be any positive number larger than $\sqrt{L}\|y_0\|_V$ and choose $T > 0$ as in (5.5.11).

From (5.3.8), we have

$$\|F\|_{[L^2(Q_T)]^3} \leq C_1\sqrt{T}, \tag{5.5.17}$$

where $C_1 = \sqrt{\text{meas}\,\Omega}\zeta$.

Select T^* so that

$$0 < T^* < \min\left\{\frac{q^2 - L\|y_0\|_V^2}{LC_1^2}, T, 1\right\}. \tag{5.5.18}$$

Now, we have

$$\|\mathbf{BFA}u\|^2_{L^2(0,T^*;V)} \leq L\|y_0\|^2_V + L\|F(\mathbf{A}u)\|^2_{(L^2(Q_{T^*}))^3}$$

$$< L\|y_0\|^2_V + LC_1^2T^* < L\|y_0\|^2_V + q^2 - L\|y_0\|^2_V = q^2.$$

Hence, **BFA** maps $\mathscr{B}_q(0)$ into itself if (5.5.18) is satisfied.

Thus, by Schauder's fixed point theorem, **BFA** has a fixed point u, which is a solution of the system (5.3.1), and which satisfies all of the requirements of Theorem 5.1. As usual, we may select ∇p in $L^2(0, T^*; G(\Omega))$ to complement the solution $u \in L^2(0, T^*; H)$ in Theorem 5.1.

This completes the proof of existence for Theorem 5.1.

5.5.3.2 Proof of Uniqueness

To prove that a solution found in Sect. 5.5.3.1 is unique, we will argue by contradiction.

Suppose that there are two different solutions

$$\left\{ z^{(1)} = (z_1^{(1)}, \ldots, z_n^{(1)}), y^{(1)}, p^{(1)} \right\}$$

and

$$\left\{ z^{(2)} = (z_1^{(2)}, \ldots, z_n^{(2)}), y^{(2)}, p^{(2)} \right\}$$

to system (5.3.1), satisfying the properties described in Theorem 5.1 on some time interval $[0, T]$. Without loss of generality, we assume that these two solutions are different right from $t = 0$.

By (5.5.12), with $z_i^{(j)}$ in place of $w_i^{(j)}$, $i = 1, \ldots, n$, and $y^{(j)}$ in place of $u^{(j)}$, both for $j = 1, 2$, we see that for any $T_0 \in (0, T]$, and each $i = 1, \ldots, n$,

$$\|z_i^{(1)} - z_i^{(2)}\|_{C([0,T];R^3)} \leq \frac{\sqrt{T \operatorname{meas}(S)}}{\operatorname{meas}(S) - Cq\sqrt{T}} \|y^{(1)} - y^{(2)}\|_{(L^2(Q_T))^3}. \qquad (5.5.19)$$

Now, for $(y^{(1)} - y^{(2)})$, we have the following initial- and boundary-value problem:

$$\frac{\partial(y^{(1)} - y^{(2)})}{\partial t} = \nu\Delta(y^{(1)} - y^{(2)}) + (F(z^{(1)}) - F(z^{(2)})) - \nabla(p^{(1)} - p^{(2)}) \quad \text{in } Q_T,$$

$$\operatorname{div}(y^{(1)} - y^{(2)}) = 0 \quad \text{in } Q_T,$$

$$(y^{(1)} - y^{(2)}) = 0 \ \text{ in } \Sigma_T,$$

$$(y^{(1)} - y^{(2)})|_{t=0} = 0.$$

According to (5.5.15), we have

$$\|y^{(1)} - y^{(2)}\|^2_{(L^2(Q_T))^3} \le L \int_0^T \int_\Omega \|F(z^{(1)}) - F(z^{(2)})\|^2_{R^3} dx dt. \tag{5.5.20}$$

In turn, similar to (5.5.16),

$$\|F(z^{(1)}) - F(z^{(2)})\|_{R^3} \le N(T) \sum_{j=1}^n \|z_j^{(1)} - z_j^{(2)}\|_{C([0,T];R^3)} \tag{5.5.21}$$

for some $N(T)$ is non-increasing at $T \to 0^+$.

Hence, combining (5.5.19)–(5.5.21) yields

$$\|y^{(1)} - y^{(2)}\|_{(L^2(Q_T))^3} \le \frac{n N(T) T \sqrt{L \operatorname{meas}(S) \operatorname{mes}(\Omega)}}{\operatorname{meas}(S) - Cq\sqrt{T}} \|y^{(1)} - y^{(2)}\|_{(L^2(Q_T))^3}. \tag{5.5.22}$$

Now, if, in addition to the above, we select T small enough so that

$$0 < T < \frac{\operatorname{meas}(S)}{4nN(T)\sqrt{L\operatorname{meas}(S)\operatorname{meas}(\Omega)}}, \tag{5.5.23}$$

we will arrive to a contradiction:

$$\|y^{(1)} - y^{(2)}\|_{(L^2(Q_T))^3} < \frac{1}{2}\|y^{(1)} - y^{(2)}\|_{(L^2(Q_T))^3}.$$

Therefore, $y^{(1)} \equiv y^{(2)}$ on $[0, T]$, and thus by (5.5.19), $z_i^{(1)} \equiv z_i^{(2)}$ for $i = 1, \ldots, n$ on $[0, T]$.

This ends the proof of Theorem 5.1.

Chapter 6
Well-Posedness of 2D or 3D Bio-Mimetic Swimmers: The Case of Navier–Stokes Equations

In this chapter we will extend the mathematical well-posedness results of Chap. 5 to the case of a fluid governed by either 2D or 3D Navier–Stokes equations, assuming that *a swimmer can engage prescribed geometric controls*. (For the case, when a swimmer does not use geometric controls, we refer to [33].)

We will use the same scheme of proofs as in Chap. 5 with the following change: we will need to modify Sect. 5.5.2.2 in Chap. 5, dealing with the regularity results for the uncoupled fluid equation, describing the continuity properties of operator **B**, along a respective lemma in [33].

6.1 Problem Setup and Main Results

6.1.1 Problem Setting

In this chapter we will consider the following initial- and boundary-value problem (compare to (5.3.1)) describing the swimmers' models from Part I in the case when the fluid is governed by the Navier–Stokes equations:

$$
\begin{cases}
u_t + (u \cdot \nabla)u = \nu \Delta u + F - \nabla p & \text{in } Q_T = (0, T) \times \Omega, \\
\operatorname{div} u = 0 & \text{in } Q_T, \\
u = 0 & \text{in } \Sigma_T = (0, T) \times \partial\Omega, \\
u(0, \cdot) = u_0 & \text{in } \Omega \subset R^K, \quad K = 2, 3, \\
\frac{dz_i}{dt} = \frac{1}{\operatorname{meas}(S)} \int_{S(z_i(t))} u(t, x)dx, & \\
\quad z_i(0) = z_{i,0}, \quad i = 1, \dots, n,
\end{cases}
\tag{6.1.1}
$$

© The Author(s), under exclusive license to Springer Nature Switzerland AG 2021
A. Khapalov, *Bio-Mimetic Swimmers in Incompressible Fluids*, Lecture Notes
in Mathematical Fluid Mechanics, https://doi.org/10.1007/978-3-030-85285-6_6

where $x = (x_1, \ldots, x_K)$, $u(t, x) = (u_1(t, x), \ldots, u_K(t, x))$,

$$z(t) = (z_1(t), \ldots, z_n(t)) \in [R^K]^n, \quad v(t) = (v_1(t), \ldots, v_{n-2}(t)) \in R^{n-2},$$

$$w(t) = (w_1(t), \ldots, w_{n-1}(t)) \in R^{n-1},$$

a set of geometric controls (see (5.2.2))

$$S_i(0, t), \quad t \in [0, T], \quad i = 1, \ldots, n, \tag{6.1.2}$$

is prescribed, and F can be of any configuration of the swimmers' internal rotational and elastic forces introduced in Chaps. 2–4 in 2D and 3D cases, see (7.2.3)–(5.3.6).

6.1.2   Main Results

Theorem 6.1 (2D Swimmer'S Model) *Let us assume that $K = 2$ and assumptions of Theorem 5.1 hold with $u_0 \in H$ and, in particular, a set of geometric controls $S_i(0, t)$, $t \in [0, T]$, $i = 1, \ldots, n$ as in (5.2.2) be given. Then, there exists a $T^* \in (0, T]$ such that system (6.1.1) and (6.1.2) admits a unique solution (u, z) in $C(0, T^*; H) \cap L^2(0, T^*; V) \times [C([0, T^*]; R^2)]^n$, $\nabla p \in G(\Omega)$ (see also Proposition 6.1 below), and restrictions (5.2.5) and (5.2.6) hold on $[0, T_*]$.*

Theorem 6.2 (3D Swimmer's Model) *Let $K = 3$ and assumptions of Theorem 6.1 hold with $u_0 \in V$. Then, for the same T^* as in Theorem 6.1, (6.1.1) admits a unique solution (u, z, p) such that*

$$u \in C([0, T^*]; V), \quad u_t, \Delta u, \nabla p \in [L^2(Q_{T^*})]^3.$$

Theorem 6.3 (Additional Regularity) *If $u_0 \in [H^2(\Omega)]^3 \cap V$, then, under assumptions of Theorem 6.2, $u \in [H^{2,1}(Q_{T^*})]^3 \cap C([0, T^*]; V)$. In turn, if $u_0 \in [H^2(\Omega)]^2 \cap V$, then $u \in [H^{2,1}(Q_{T^*})]^2 \cap C([0, T^*]; V)$ under assumptions of Theorem 6.1.*

This result is an immediate consequence of Theorem 6.6.

Similar results for the case when geometric controls are not engage can be found in [33].

6.2   Proofs of the Main Results

In Sect. 5.5.2, we established that, for sufficiently small $T^* > 0$, namely, satisfying (5.5.11), the operators

$$\mathbf{A} : \mathscr{B}_q(0) \longrightarrow [C([0, T]; R^3)]^n, \quad \mathbf{A}u = w = (w_1, \ldots, w_n),$$

$$\mathscr{B}_q(0) = \{\phi \in \mathscr{H} \,|\, \|\phi\|_{\mathscr{H}} \leq q\}, \quad \|\phi\|_{\mathscr{H}} = \sqrt{\|\phi\|^2_{L^2(0,T;(L^2(\Omega))^3)} + \|\phi\|^2_{L^2(0,T;V)}},$$

$$\mathbf{F} : [C([0,T]; R^3)]^n \longrightarrow (L^2(Q_T))^3, \quad \mathbf{F}p = F(p),$$

$$\mathbf{B} : (L^2(Q_T))^3 \longrightarrow L^3(0,T;V) \bigcap L^2(0,T;H), \quad \mathbf{B}f = y$$

are all continuous, while **A** is also compact. Respectively, the operator

$$\mathbf{BFA} : \ B_q(0) \longrightarrow L^2(0,T;V) \bigcap L^2(0,T;H), \quad \mathbf{BFA}u = y$$

is both continuous and compact.

In this chapter, we will deal with a different operator $\mathbf{B}_{NS}$, defined in the next subsection.

6.2.1   Solution Mapping for Decoupled Navier–Stokes Equations

Let $T > 0$ be as in (5.5.11). Consider the following decoupled initial- and boundary-value problem for the non-stationary Navier–Stokes equations:

$$u_t + (u \cdot \nabla)u = \nu \Delta u + f(x,t) - \nabla p \ \text{ in } \ Q_T,$$

$$\text{div } u = 0 \ \text{ in } \ Q_T, \quad u = 0 \ \text{ in } \ \Sigma_T, \quad u|_{t=0} = u_0 \in H. \tag{6.2.1}$$

To prove Theorem 6.1 and 6.2 along the scheme of Chap. 5, we need to show that, for some $T^* < T$, the mapping

$$\mathbf{B}_{NS} : [L^2(Q_{T^*})]^K) \to L^2(0,T^*;V), \quad \mathbf{B}_{NS}f = u, \tag{6.2.2}$$

where u is the solution to uncoupled Navier–Stokes equations (6.1.1), is continuous, and for the same T^*,

$$\mathbf{B}_{NS}\mathbf{FT}\mathscr{B}_q \ \longrightarrow \ \mathscr{B}_q.$$

6.2.2   Preliminary Results

Let us recall some classical results relevant to the argument below.

- Consider the following trilinear continuous map on $[H_0^1(\Omega)]^K \times [H_0^1(\Omega)]^K \times [H_0^1(\Omega)]^K$:

$$b(u, v, w) := \sum_{i,j=1}^{K} \int_{\Omega} u_i \, (d_i v_j) \, w_j \, dx, \tag{6.2.3}$$

where D_j is the differentiation operator with respect to x_j. Due to Lemma 3.4 (page 292) in [56, Chapter III], for $K = 2$, for all $u, v, w \in [H_0^1(\Omega)]^2$, we have

$$|b(u, v, w)| \leq \sqrt{2} \left(\|u\|_{H_0^1} \|u\|_{L^2} \right)^{\frac{1}{2}} \|v\|_{H_0^1} \left(\|w\|_{H_0^1} \|w\|_{L^2} \right)^{\frac{1}{2}}. \tag{6.2.4}$$

- **Theorem 6.4 (Theorems III.3.1 and III.3.2 in [56, p. 282 and p. 294])** *Let T be any positive number, $K = 2$, $f \in L^2(0, T ; [L^2(\Omega)]^2)$ and $u_0 \in H$. Then, there exists a unique solution $u \in C([0, T]; H) \cap L^2(0, T ; V)$ of (6.1.1) on $[0, T]$ and the following estimate holds (see estimates (45) and (48) in Theorem 11 [36, pp. 170–171] as well as our estimates (6.2.15)–(6.2.17) when $u_1 = u$ and $u_2, f_2 = 0$):*

$$\|u\|_{C(0,T;H)} + \|u\|_{L^2(0,T;V)} \leq L^* \left(\|u_0\|_{L^2} + \|f\|_{L^2(0,T;[L^2(\Omega)]^2)} \right) \tag{6.2.5}$$

for some positive constant L^*.
- **Proposition 6.1 (Theorem V.1.7.1 in [52, p. 295])** *Under the assumptions of Theorem 6.4, there exists a function $\mathscr{P} \in L^2(0, T ; L^2(\Omega))$ such that $p = \mathscr{P}_t$ is an associated pressure of u and*

$$\nabla p = f - u_t + \nu \Delta u - (u \cdot \nabla) u$$

in the sense of distributions in $(0, T) \times \Omega$.
- **Theorem 6.5 (Theorem 9, [37, p. 203 and Lemma 9, p. 194])** *Let $T > 0$ be any given positive number, $K = 3$, $f \in L^2(0, T ; [L^2(\Omega)]^3)$, $u_0 \in V$. Then for any T_1 satisfying*

$$T_1 \in (0, \mathscr{T}), \quad \mathscr{T} = \min \left\{ T, C^o \left[\|u_0\|_{H_0^1}^2 + \|f\|_{L^2(0,T;[L^2(\Omega)]^3)}^2 \right]^{-2} \right\} \tag{6.2.6}$$

(C^o depends on Ω and ν only), (6.1.1) admits a unique solution (u, p) on $(0, T_1)$ such that $u_t, \Delta u, \nabla p \in [L^2(Q_{T_1})]^3$ and $u \in C([0, T_1]; V)$. Furthermore, the following estimates hold (see (6.2.21) and (6.2.23) when $u_1 = u$ and $u_2, f_2 = 0$):

$$\|u\|_{C([0,T_1];H)} + \|u\|_{L^2(0,T_1;V)} \leq L^{**} \left(\|u_0\|_{L^2} + \|f\|_{L^2(0,T_1;[L^2(\Omega)]^3)} \right) \tag{6.2.7}$$

for some positive constant $L^{**} > 0$ and (see [37, Lemma 9, p. 194, (55)]):

$$\|u\|_{C([0,T_1];V)} \leq \hat{L} \left(\|u_0\|_{H_0^1} + \|f\|_{L^2(0,T_1;[L^2(\Omega)]^3)} \right) \tag{6.2.8}$$

for some positive constant $\hat{L} > 0$.

- **Theorem 6.6 (Additional Regularity, [36, Theorem 17, pp. 183–184])** *If $u_0 \in [H^2(\Omega)]^3 \cap V$, then, under assumptions of Theorem 6.5,*

$$u \in [H^{2,1}(Q_{T_1})]^3 \cap C([0, T_1]; V).$$

In turn, if $u_0 \in [H^2(\Omega)]^2 \cap V$, then $u \in [H^{2,1}(Q_{T_1})]^2 \cap C([0, T_1]; V)$ under assumptions of Theorem 6.4.

Remark 6.1 (Duration of Solutions to Navier–Stokes Equations)

- Let us note that, for spatial dimension $K = 2$, Theorem 6.4 does not restrict the duration of solutions to the (decoupled) initial- and boundary-value problem for the Navier–Stokes equation.
- In turn, for $K = 3$, Theorem 6.5 has such restriction which depends on u_0 and f as given in (6.2.6).
- For $K = 3$, we will need to identify a time interval, denote it here by $(0, T_1)$, for the coupled system (6.1.1), which, given u_0, will work uniformly for any selection of f used in the proofs of Theorems 6.1 and 6.2 below; namely, we set

$$T_1 \in (0, \min\left\{T, C^o\left[||u_0||_{H_0^1}^2 + T(\text{meas})(S)\zeta^2\right]^{-2}\right\}) \quad K = 3, \qquad (6.2.9)$$

where ζ is from (5.3.8).

6.2.3   Continuity of $\mathbf{B}_{NS}$

The following lemma follows the guidelines of Lemma 4.4 in work [33] (which does not consider geometric controls).

Lemma 6.1 (Continuity of $\mathbf{B}_{NS}$) *Select any $T^* \in (0, T_1)$. Then,*

$$\|\mathbf{S}f\|_{L^2(0,T^*;V)} \leq L_* \left(\|u_0\|_{L^2} + \|f\|_{L^2(0,T^*;[L^2(\Omega)]^K)}\right), \quad K = 2, 3, \qquad (6.2.10)$$

and for $K = 2$,

$$\|\mathbf{B}_{NS}f - \mathbf{B}_{NS}g\|_{L^2(0,T^*;V)}$$

$$\leq L'e^{C'\max\{\|f\|_{L^2(0,T^*;[L^2(\Omega)]^2)}^2, \|g\|_{L^2(0,T^*;[L^2(\Omega)]^2)}^2\}} \|f - g\|_{L^2(0,T^*;[L^2(\Omega)]^2)}, \qquad (6.2.11)$$

while for $K = 3$,

$$\|\mathbf{B}_{NS}f - \mathbf{B}_{NS}g\|_{L^2(0,T^*;V)}$$

$$\leq L'' \, e^{C'' \, \max\{\|f\|^8_{L^2(0,T^*;[L^2(\Omega)]^3)}, \|g\|^8_{L^2(0,T^*;[L^2(\Omega)]^3)}\}} \, \|f - g\|_{L^2(0,T^*;[L^2(\Omega)]^3)} \,,$$

$$(6.2.12)$$

where $L_*, L', L'', C',$ *and* C'' *are some positive constants.*

Proof of the Boundedness of $\mathbf{B}_{NS}$ For any $f \in L^2(0, T^* ; [L^2(\Omega)]^K)$, estimate
(6.2.10) follows immediately from (6.2.5) in the 2D case and from (6.2.7) in the 3D
case.

Proof of the Continuity of $\mathbf{B}_{NS}$

Step 1 Let $K = 2$ and $(u_1 = \mathbf{B}_{NS}f_1, u_2 = \mathbf{B}_{NS}f_2)$ be a pair of solutions of
(6.1.1) with the forcing terms, respectively, f_1 and f_2 from $L^2(0, T^*; [L^2(\Omega)]^2)$
(u_0 is fixed).

To derive (6.2.11), it is sufficient to establish the following estimate:

$$\|u_1 - u_2\|^2_{L^2(0,T^*;V)}$$

$$\leq \frac{4 \, C^2_\Omega}{\nu^2} \, e^{\frac{8}{\nu} \, \max\{\|u_1\|^2_{L^2(0,T^*;V)}, \|u_2\|^2_{L^2(0,T^*;V)}\}} \, \|f_1 - f_2\|^2_{L^2(0,T^*;[L^2(\Omega)]^2)} \,,$$

$$(6.2.13)$$

where C_Ω is the constant from the Poincaré inequality.

Step 2 From the proof of Theorem III.3.2 in [56], we obtain that for any $t \in (0, T^*)$, the difference $u_1(t) - u_2(t)$ satisfies the following equality (see in particular
(3.65) in [56, Chapter III] or (5), page 145 in [37]):

$$\frac{d}{dt} \left(\frac{1}{2} \|u_1(t) - u_2(t)\|^2_{L^2} \right) + \nu \, \|u_1(t) - u_2(t)\|^2_{H^1_0}$$

$$= \langle f_1(t) - f_2(t), u_1(t) - u_2(t) \rangle_{L^2}$$

$$- b\big(u_1(t) - u_2(t), u_2(t), u_1(t) - u_2(t)\big) ,$$

where $\langle \cdot, \cdot \rangle_{L^2}$ stands for the scalar product in $L^2(\Omega)$. Hence, we have

$$\frac{d}{dt} \left(\|u_1(t) - u_2(t)\|^2_{L^2} \right) + 2\nu \, \|u_1(t) - u_2(t)\|^2_{H^1_0} \qquad\qquad (6.2.14)$$

$$\leq 2 \, \|f_1(t) - f_2(t)\|_{L^2} \, \|u_1(t) - u_2(t)\|_{L^2}$$

$$+ 2 \left| b\big(u_1(t) - u_2(t), u_2(t), u_1(t) - u_2(t)\big) \right|$$

$$\le 2\,C_\Omega \, \|f_1(t) - f_2(t)\|_{L^2} \, \|u_1(t) - u_2(t)\|_{H_0^1}$$

$$+ 2\sqrt{2} \, \|u_2(t)\|_{H_0^1} \, \|u_1(t) - u_2(t)\|_{L^2} \, \|u_1(t) - u_2(t)\|_{H_0^1} \,,$$

$$\le \frac{2\,C_\Omega^2}{\nu} \, \|f_1(t) - f_2(t)\|_{L^2}^2 + \frac{\nu}{2} \, \|u_1(t) - u_2(t)\|_{H_0^1}^2$$

$$+ \frac{4}{\nu} \, \|u_2(t)\|_{H_0^1}^2 \, \|u_1(t) - u_2(t)\|_{L^2}^2 + \frac{\nu}{2} \, \|u_1(t) - u_2(t)\|_{H_0^1}^2 \,,$$

where, to evaluate $|b(\cdot, \cdot, \cdot)|$, we used (6.2.4). Therefore,

$$\frac{d}{dt} \left(\|u_1(t) - u_2(t)\|_{L^2}^2 \right) + \nu \, \|u_1(t) - u_2(t)\|_{H_0^1}^2$$

$$\le \frac{2\,C_\Omega^2}{\nu} \, \|f_1(t) - f_2(t)\|_{L^2}^2$$

$$+ \frac{4}{\nu} \, \|u_2(t)\|_{H_0^1}^2 \, \|u_1(t) - u_2(t)\|_{L^2}^2 \,. \qquad (6.2.15)$$

In turn, (6.2.15) implies that

$$\frac{d}{dt} \left(\|u_1(t) - u_2(t)\|_{L^2}^2 \right) \le \frac{2\,C_\Omega^2}{\nu} \, \|f_1(t) - f_2(t)\|_{L^2}^2 + \frac{4}{\nu} \, \|u_2(t)\|_{H_0^1}^2 \, \|u_1(t) - u_2(t)\|_{L^2}^2 \,.$$

Now, we can apply Gronwall's inequality to arrive at

$$\|u_1(t) - u_2(t)\|_{L^2}^2 \le \frac{2\,C_\Omega^2}{\nu} \int_0^t \|f_1(s) - f_2(s)\|_{L^2}^2 \, ds \cdot \exp\left(\frac{4}{\nu} \|u_2\|_{L^2(0,t;V)}^2 \right),$$

$$t \in [0, T^*]. \qquad (6.2.16)$$

In turn, integrating (6.2.15), we deduce from (6.2.16) that

$$\nu \|u_1 - u_2\|_{L^2(0,T^*;V)}^2 \le \|u_1(T^*) - u_2(T^*)\|_{L^2}^2 + \nu \|u_1 - u_2\|_{L^2(0,T^*;V)}^2$$

$$\le \frac{2\,C_\Omega^2}{\nu} \int_0^t \|f_1(s) - f_2(s)\|_{L^2}^2 \, ds + \max_{s \in [0,T^*]} \|u_1(s) - u_2(s)\|_{L^2}^2 \cdot \frac{4}{\nu} \|u_2\|_{L^2(0,T^*;V)}^2$$

$$\le \frac{2\,C_\Omega^2}{\nu} \int_0^t \|f_1(s) - f_2(s)\|_{L^2}^2 \, ds \left(1 + \frac{4}{\nu} \|u_2\|_{L^2(0,T^*;V)}^2 \, e^{\frac{4}{\nu} \|u_2\|_{L^2(0,T^*;V)}^2} \right).$$

From here, making use of estimate $(1 + xe^x) < 2e^{2x}$ for $x \geq 0$, we arrive at

$$\|u_1 - u_2\|^2_{L^2(0,T^*;V)} \leq \frac{4\,C^2_\Omega}{\nu^2} e^{\frac{8}{\nu}\|u_2\|^2_{L^2(0,T^*;V)}} \|f_1 - f_2\|^2_{L^2(0,T^*;[L^2(\Omega)]^2))},$$
(6.2.17)

implying (6.2.11). This completes the proof for $K = 2$.

Remark 6.2 An estimate similar to (6.2.13) can also be found in [36], see Theorem 11 and estimates (45) and (48) on pp. 170–171.

Step 3: The Case $K = 3$ Consider any two solutions $u_i = \mathbf{B}_{NS} f_i, i = 1, 2$, to (6.1.1) for some f_1, f_2 in $L^2(0, T^*; H)$.

Combining estimate (11) from [36, p. 147], namely,

$$\|\psi\|_{L^4(\Omega)} \leq D^* \|\psi\|_{H^1_0} \quad \forall \psi \in H^1_0(\Omega), \text{ where } D^* \text{ is some positive constant,}$$

with the fact that $u_2 \in C([0, T^*]; V)$ (see Theorem 6.5) allows us to define

$$C_* = \max_{t \in [0,T^*]} |u_2(t)|_{L^4} \leq 3^{1/4} D^* \|u_2\|_{C([0,T^*];V)},$$
(6.2.18)

where

$$|u_2(t)|_{L^4} = \left[\sum_{i=1}^{3} \int_\Omega u^4_{2i}(t, x) dx \right]^{1/4}, \quad u_2 = (u_{21}, u_{22}, u_{23}).$$

To evaluate $b(u_1(t) - u_2(t), u_2(t), u_1(t) - u_2(t))$ in the 3D case, we will use the chain of estimates from [36, p. 145] as follows:

$$\left| b(u_1(t) - u_2(t), u_2(t), u_1(t) - u_2(t)) \right|$$

$$\leq \sqrt{3} |u_2(t)|_{L^4} \|u_1(t) - u_2(t)\|_{H^1_0} |u_1(t) - u_2(t)|_{L^4}$$

$$\leq \sqrt{3} C_2 \|u_1(t) - u_2(t)\|_{H^1_0} |u_1(t) - u_2(t)|_{L^4}.$$

Then, making use of estimate (12) from [36, p. 147 and (5) on p. 10], namely,

$$\|\psi\|_{L^4(\Omega)} \leq \delta \|\psi\|_{H^1_0(\Omega)} + \frac{3^{3/4}}{\delta^3} \|\psi\|_{L^2(\Omega)}, \quad \psi \in H^1_0(\Omega), \ \delta > 0,$$

we obtain as in [36, p. 145]:

$$\left| b(u_1(t) - u_2(t), u_2(t), u_1(t) - u_2(t)) \right|$$

$$\leq \ \sqrt{3}C_* \|u_1(t) - u_2(t)\|_{H_0^1} \left[\varepsilon \|u_1(t) - u_2(t)\|_{H_0^1} + C_\varepsilon \|u_1(t) - u_2(t)\|_{[L^2(\Omega)]^3} \right],$$

$$\leq \ 2\sqrt{3}C_*\varepsilon \|u_1(t) - u_2(t)\|_{H_0^1}^2 + \sqrt{3}C_* \frac{C_\varepsilon^2}{4\varepsilon} \|u_1(t) - u_2(t)\|_{[L^2(\Omega)]^3}^2, \quad \varepsilon > 0,$$

where $C_\varepsilon = 3(3)^{3/4}/\varepsilon^3$.

Select now $\varepsilon = (2\sqrt{3}C_2\})^{-1}(\nu/4)$; then,

$$\left| b\big(u_1(t) - u_2(t), u_2(t), u_1(t) - u_2(t)\big) \right|$$

$$\leq \ \frac{\nu}{4} \|u_1(t) - u_2(t)\|_{H_0^1}^2 + \frac{6C_2^2 C_\varepsilon^2}{\nu} \|u_1(t) - u_2(t)\|_{[L^2(\Omega)]^3}^2,$$

$$\leq \ \frac{\nu}{4} \|u_1(t) - u_2(t)\|_{H_0^1}^2 + D_* \|u_1(t) - u_2(t)\|_{[L^2(\Omega)]^3}^2, \tag{6.2.19}$$

where we can set, due to (6.2.8) and (6.2.18),

$$D_* \ = \hat{C} \hat{L}_S^8 \left(\|u_0\|_V + \|f_2\|_{L^2(0,T_1\,;[L^2(\Omega)]^3)} \right)^8, \tag{6.2.20}$$

and $\hat{C}$ depends on ν. Similar to (6.2.14), we obtain

$$\frac{d}{dt} \left(\|u_1(t) - u_2(t)\|_{L^2}^2 \right) + 2\nu \|u_1(t) - u_2(t)\|_{H_0^1}^2 \tag{6.2.21}$$

$$\leq 2 \|f_1(t) - f_2(t)\|_{L^2} \|u_1(t) - u_2(t)\|_{L^2} + 2 \left| b\big(u_1(t) - u_2(t), u_2(t), u_1(t) - u_2(t)\big) \right|$$

$$\leq \ \frac{2C_\Omega^2}{\nu} \|f_1(t) - f_2(t)\|_{L^2}^2 + \frac{\nu}{2} \|u_1(t) - u_2(t)\|_{H_0^1}^2$$

$$+ \ \frac{\nu}{2} \|(u_1(t) - u_2(t))\|_{H_0^1}^2 + 2D_* \|(u_1(t) - u_2(t))\|_{[L^2(\Omega)]^3}^2 .$$

Instead of (6.2.15), we have

$$\frac{d}{dt} \left(\|u_1(t) - u_2(t)\|_{L^2}^2 \right) + \nu \|u_1(t) - u_2(t)\|_{H_0^1}^2$$

$$\leq \ \frac{2C_\Omega^2}{\nu} \|f_1(t) - f_2(t)\|_{L^2}^2 + 2D_* \|(u_1(t) - u_2(t))\|_{[L^2(\Omega)]^3}^2 ,$$

$$\tag{6.2.22}$$

and, furthermore, in place of (6.2.17),

$$\|u_1 - u_2\|^2_{L^2(0,T^*;V)} \leq \frac{4\,C^2_\Omega}{\nu^2}\, e^{4D_*T^*}\, \|f_1 - f_2\|^2_{L^2(0,T^*;[L^2(\Omega)]^2))}\,, \qquad (6.2.23)$$

implying (6.2.12) in view of (6.2.20). This completes the proof of Lemma 6.1.

6.2.4   Proof of Theorems 6.1 and 6.2

Existence   From (6.2.10), we obtain

$$\|\mathbf{S}f\|^2_{L^2(0,T^*;V)} \leq 2L^2_* \left(\|u_0\|^2_{L^2} + \|f\|^2_{L^2(0,T^*;[L^2(\Omega)]^K)} \right) \qquad (6.2.24)$$

Respectively, the existence follows by the argument in Sect. 5.5.3.1 with $2L^2_*$ in place of L from (5.5.15) and

$$0 < T^* < \min \left\{ \frac{q^2 - 2L^2_*1\|y_0\|^2_V}{2L^2_*C^2_1}, T, T_1, 1 \right\}, \qquad (6.2.25)$$

where T_1 is from Remark 6.1 and $T > 0$ is from (5.5.11).

Proof of Uniqueness   We can prove uniqueness along the scheme of Sect. 5.5.3.2, with the help of Lemma 6.1 in place of (5.5.15) (see a more detailed argument in work [33], dealing with the case of swimmers with no geometric controls engaged).

Part III
Micromotions and Local Controllability for Bio-Mimetic Swimmers in 2D and 3D Incompressible Fluids

The subject of our interest in Chaps. 7–8 is the swimming phenomenon from the viewpoint of mathematical controllability theory for partial differential equations in its local aspect. More precisely, in Chaps. 7–8, we will focus on the steering capability of a swimmer near its initial position, when it makes use of its internal control forces only, while assuming that its geometric controls are fixed.

Chapter 7
Local Controllability of 2D and 3D Swimmers: The Case of Non-stationary Stokes Equations

In this chapter we will discuss local controllability (or steering) properties of swimmers in a fluid described by the non-stationary Stokes equations, with focus on the 2D case. Our approach is based on the finite-dimensional (in R^K, $K = 2, 3$) Inverse Function Theorem and makes use of implicit (generalized) Fourier series representation for solutions to these linear fluid equations. In Chap. 8 we will generalize this approach to the nonlinear Navier–Stokes equations, while avoiding the aforementioned series representations.

7.1 Definitions of Controllability for Bio-Mimetic Swimmers

The following definitions were introduced in the work [20] (see also [25, Chapter 14], for the case when the swimmer does not engage its geometric controls. In this monograph, we will modify them for the case of active geometric controls.

Let a set of geometric controls, as in (5.2.2), be given.

For a given initial state

$$\{u_0, z(0)\} = \{u_0, \ z_i(0), \ i = 1, \dots, n\}, \tag{7.1.1}$$

denote by

$$\{u^*(t, x), \ z_*(t)\}, \ \ t > 0, \tag{7.1.2}$$

the solution pair to any of the systems (5.3.1) and (5.3.2) or (6.1.1) and (6.1.2), generated by the zero controls $v_i = 0$, $i = 1, \dots, n-2$; $w_j = 0, j = 1, \dots, n-1$.

Assuming that the elastic Hooke's forces are present in a model at hand, let us consider the following *initial equilibrium state/datum*:

© The Author(s), under exclusive license to Springer Nature Switzerland AG 2021
A. Khapalov, *Bio-Mimetic Swimmers in Incompressible Fluids*, Lecture Notes
in Mathematical Fluid Mechanics, https://doi.org/10.1007/978-3-030-85285-6_7

$$\{\{u_0, z(0)\} = \{u_0 = 0, \quad z_i(0), \quad i = 1, \ldots, n\} \mid l_{i-1} = \| z_i(0) - z_{i-1}(0) \|_{R^2}, \quad i = 2, \ldots, n\}. \tag{7.1.3}$$

In this case, the fluid and the swimmer do not move for any $t > 0$, if no internal control forces are engaged,

$$u^*(t, x) \equiv 0, \quad z_*(t) = z(0), \quad t \geq 0. \tag{7.1.4}$$

Local Controllability Near Equilibrium for Individual Points $z_i(t)$'s
We ask

- *Given an initial equilibrium datum (7.1.3) and a pre-assigned moment $T > 0$, can we move a given z_i to any desirable position $z_i(T)$ at time $t = T$ within some neighborhood of $z_i(0)$, by engaging controls v_i's and/or w_i's?*

This question poses, what we call, *the local controllability problem* with respect to z_i near equilibrium at time T.

Local Controllability Near Equilibrium for a Swimmer
However, the question of main interest, associated with the actual motion of a swimmer, is the following:

- *Given an equilibrium initial datum (7.1.3), can we move the center of mass of our swimmer, namely the point*

$$z_c(t) = \frac{1}{n} \sum_{i=1}^{n} z_i(t),$$

anywhere within some neighborhood of its initial equilibrium position

$$z_c(0) = \frac{1}{n} \sum_{i=1}^{n} z_i(0)$$

at some pre-assigned moment $T > 0$?

This question poses, what we call, the *local controllability problem with respect to $z_c(0)$ near equilibrium* at time T.

Local Controllability Near a Drifting Trajectory with Active Geometric Controls
For a given set of geometric controls, consider now, what we can call, *the swimmer's equilibrium initial condition with active geometric controls,*

$$\{\{u_0, z(0)\} = \{u_0, \quad z_i(0), \quad i = 1, \ldots, n\} \mid l_{i-1} = \| z_i(0) - z_{i-1}(0) \|_{R^2}, \quad i = 2, \ldots, n\}, \tag{7.1.5}$$

when u_0 does not have to be zero, but the swimmer's Hooke's forces (if such are present in a model), are equal to zero. Then, assuming that no internal forces are engaged for $t > 0$, we will call the respective solution pair,

$$\{u_d(t, x), \ z_d(t)\}, \quad t > 0, \tag{7.1.6}$$

a *drifting solution with active geometric controls or just a drifting trajectory*, while $z_d(t) = (z_{1d}(t), \ldots, z_{nd}(t))$ will be a *drifting trajectory of a swimmer*.

We ask

Given the initial datum in (7.1.5), can we move a desirable point z_i of the swimmer or its center of mass z_c anywhere within some neighborhood, respectively, of $z_{id}(T)$ or of

$$z_{cd}(T) = \frac{1}{n} \sum_{i=1}^{n} z_{id}(T), \tag{7.1.7}$$

for some pre-assigned moment $T > 0$?

We will call the problems posed by these two questions—the *local controllability problems near a drifting trajectory*.

7.2 Main Results

We will consider a version of 2D swimmer's model from Chap. 5:

$$\begin{cases} u_t = \nu \Delta u + F - \nabla p & \text{in } Q_T = (0, T) \rtimes \Omega, \\ \text{div } u = 0 & \text{in } Q_T, \\ u = 0 & \text{in } \Sigma_T = (0, T) \times \partial\Omega, \\ u(0, \cdot) = u_0 \in V & \text{in } \Omega \subset R^2, \\ \frac{dz_i}{dt} = \frac{1}{\text{meas}(S)} \int_{S(z_i(t))} u(t, x) dx, \ z_i(0) = z_{i,0}, \ i = 1, \ldots, n, \end{cases} \tag{7.2.1}$$

where $x = (x_1, x_2)$, $u(t, x) = (u_1(t, x), u_2(t, x))$,

$$z(t) = (z_1(t), \ldots, z_n(t)) \in [R^2]^n, \quad v(t) = (v_1(t), \ldots, v_{n-2}(t)) \in R^{n-2},$$

$$w(t) = (w_1(t), \ldots, w_{n-1}(t)) \in R^{n-1},$$

a set of geometric controls (as in (5.2.2))

$$S_i(0, t), \quad t \in [0, T], \quad i = 1, \ldots, n, \tag{7.2.2}$$

is prescribed, and F can be of any configuration of the following swimmers' internal *control* rotational and elastic forces (no Hooke's uncontrolled forces, see Remark 7.1), introduced in Chap. 2, namely:

$$F_{rot2D}(t, x) = \sum_{i=2}^{n-1} v_{i-1}(t) \left[\xi_{i-1}(t, x) A(z_{i-1}(t) - z_i(t)) \right.$$

$$-\xi_{i+1}(t,x)\frac{\|z_{i-1}(t)-z_i(t)\|^2}{\|z_{i+1}(t)-z_i(t)\|^2}A(z_{i+1}(t)-z_i(t))\Bigg]$$

$$+\sum_{i=2}^{n-1}\xi_i(t,x)v_{i-1}(t)\left[A(z_i(t)-z_{i-1}(t))-\frac{\|z_{i-1}(t)-z_i(t)\|^2}{\|z_{i+1}(t)-z_i(t)\|^2}A(z_i(t)-z_{i+1}(t))\right],$$

$$\tag{7.2.3}$$

$$F_{cef2D}(t,x) = \sum_{i=2}^{n}[\xi_{i-1}(t,x)w_{i-1}(z_i(t)-z_{i-1}(t))+\xi_i(t,x)w_{i-1}(z_{i-1}(t)-z_i(t))].$$

$$\tag{7.2.4}$$

Remark 7.1 In [20] (see also [25, Ch. 14]) we considered a local controllability problem for a different setup of swimmer's models, where Hooke's uncontrolled elastic forces and a different type of rotational control forces were active only, with no controlled elastic forces, nor geometric controls engaged.

In this chapter, we will impose the following additional assumptions on the sets $S(z_i(t))$'s, besides the assumptions in Sect. 5.2.

Assumption 7.1 (See an Illustrating Fig. 7.1) *We assume, throughout Chap. 7, that a swimmer can engage prescribed geometric controls such that, in addition to assumptions in Sect. 5.2:*

•

$$S(z_i(t)) = \{x \mid -r + z_{i,1}(t) < x_1 < r + z_{i,1}(t),$$

$$\alpha_i(t, x_1 - z_{i,1}(t)) + z_{i,2}(t) < x_2 < \beta_i(t, x_1 - z_{i,1}(t)) + z_{i,2}(t)\},$$

$$z_i(t) = (z_{i,1}(t), z_{i,2}(t)), \quad t > 0, \quad i = 1, \ldots, n, \tag{7.2.5}$$

• *where $\alpha_i(t, x_1), \beta_i(t, x_1), \alpha_{ix_1}(t, x_1), \beta_{ix_1}(t, x_1), t \in [0, T], x_1 \in [-r, +r]$, are continuous functions in $[0, T] \times [-r, +r]$, except, possibly, on finitely many continuous curves of the form $x_1 = x_1(t), t \in [0, T]$ within $[0, T] \times [-r, +r]$.*
• *It is also assumed that $\alpha_i(t, x_1), \beta_i(t, x_1), \alpha_{ix_1}(t, x_1), \beta_{ix_1}(t, x_1)$ can be continuously extended to the closure of each of the parts of $[0, T] \times [-r, +r]$ separated by the aforementioned discontinuity curves.*

•

$$\int_{S_i(0,t)\Delta S_i(0,0)} dx = \hat{s}(t) \to 0 \text{ as } t \to 0, \quad i = 1, \ldots, n, \tag{7.2.6}$$

where the symbol Δ stands for the set symmetric difference operation.

A simple example when Assumption 7.1 holds is the case when geometric controls do not apply and, thus, the sets $S(z_i(t))$'s do not change their spatial

Fig. 7.1 Demonstration of rotation of a rectangular-shaped $S(z_i(t))$ when $z_i(t))$ remains static: Assumption 7.1 holds for every t

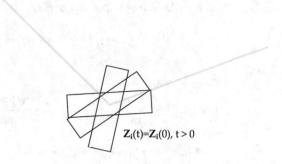

$$Z_i(t) = Z_i(0), \ t > 0$$

orientation. Alternatively, such an example can be a rotating rectangle as on Fig. 7.1, when it rotates without passing through horizontal or vertical positions. However, if, say, it rotates passing the horizontal position, then Assumption 7.1 will hold near this position if we rotate the system of spatial coordinates.

Assumption 7.2 *In this chapter we will also use the following technical assumptions:*

- *The values of $\|v_i\|_{L^\infty(0,T)}$'s and of $\|w_i\|_{L^\infty(0,T)}$'s are bounded by some positive number K.*
- *Furthermore, we assume that we will use only the constant functions for v_i's and w_k's:*

$$v_i(t) \equiv L_i, \quad w_k \equiv K_k, \quad L_i, K_i \in \{c \in R | \ |c| \le \hat{K}\} \ \ i = 1, \ldots, n-2, k = 1, \ldots, n-1.$$
$$(7.2.7)$$

- *T is selected to satisfy Theorem 5.1 in the 2D-case to ensure the well-posedness of system (7.2.1)–(7.2.4) in $[0, T]$, for the given aforementioned constant $\hat{K}$.*

7.2.1   Main Results

The following result describes momentary motions of a swimmer near its initial position.

Theorem 7.1 (Micromotions) *Let $[0, T]$ be given and Assumptions 7.1 and 7.2 hold. Then we have the following formula for micromotions of a swimmer in (7.2.1)–(7.2.4):*

$$\frac{dz_i(t)}{dt} = \frac{1}{\text{meas}\{S\}} \int_{S(z_i(0))} u_0 dx + \frac{t}{\text{meas}\{S\}} \int_{S(z_i(t_0))} P_H F(0, \cdot) dx$$

$$+ \ p(t) + (n-2)(n-1)\hat{K} o(t), \quad i = 1, \ldots, n, \tag{7.2.8}$$

where P_H is a projection operator in $(L^2(\Omega))^2$ onto H, $\hat{K}$ is from Assumption 7.2,

$$o(t)/t \to 0 \quad \text{as} \quad t \to 0+,$$

and

$$\|p(t)\|_{R^2} \leq \frac{\mu(t, u_0)}{\text{meas}^{1/2}\{S\}} + \frac{2\hat{s}^{1/2}(t) \, \| \, u_0 \, \|_{(L^2(\Omega))^2}}{\text{meas}\{S\}}, \qquad (7.2.9)$$

where, in turn, $\hat{s}(t)$ is from Assumption 7.1, $\mu(t, u_0) \to 0$ as $t \to 0+$, $0 \leq \mu(t, u_0) \leq \| u_0 \|_{(L^2(\Omega))^2}$.

Remark 7.2 (Discussion of Theorem 7.1) If control parameters v_i, w_k' in Assumption 7.2 of order t are applied (for small t), then *the directions at which each of the points $z_i(t)$, $i = 1, \ldots, n$ will move from its initial position*
  – thus, defining the motion of the swimmer's center of mass –
are *primarily determined by the projections of the swimmer's forces on the fluid velocity space at $t = 0$, averaged over their respective supports $S_i(z_i(t))$, $i = 1, \ldots, n$.*

Our next results deal with the concepts of local controllability.

Proposition 7.1 (Local Controllability of z_i's) *Let Assumptions 7.1 and 7.2 hold. Assume that in (7.2.1)–(7.2.4) only two distinct control forces are active, say, defined by v_j and v_l (or v_j and w_l, or w_j and w_l), while all other v_k's and w_m's are zeros. Assume further that v_j and v_l are independent of time. Then, if for some $i \in \{1, \ldots, n\}$ there exists a $T > 0$ such that the matrix (or the respective matrices associated with the aforementioned alternative pairs $\{v_j, w_l\}$ or $\{w_j, w_l\}$)*

$$\left(\frac{dz_i(T)}{dv_j} \Big|_{v_j, v_l = 0}, \frac{dz_i(T)}{dv_l} \Big|_{v_j, v_l = 0} \right) \qquad (7.2.10)$$

is non-degenerate, then the system (7.2.1)–(7.2.4) is locally controllable near its drifting position $z_{id}(T)$ in (7.1.6). Namely, there exists an $\varepsilon > 0$ such that

$$B_\varepsilon(z_{id}(T)) \subset \{z_i(T) \mid v_j, v_l \in R, \; v_k = 0 \text{ for } k = 1, \ldots, n-2, k \neq j, l;$$

$$w_m = 0, m = 1, \ldots, n-1\}. \qquad (7.2.11)$$

(For alternative pairs $\{v_j, w_l\}$ or $\{w_j, w_l\}$, (7.2.11) should be, respectively, modified.) In particular, for the initial equilibrium position (7.1.3) condition (7.2.12) implies the local controllability with respect to z_i near equilibrium at time T.

In other words, (7.2.11) means that the set of all possible positions of $z_i(T)$ when controls v_i's run over R will include some ε-neighborhood of $z_{id}(T)$.

Remark 7.3 In the above and below, the subscript $|_{v_j, v_l = 0}$ in $\frac{dz_i(T)}{dv_j}$ $|_{v_j, v_l = 0}$ indicates that we calculate the partial derivative with respect to v_j at $v_j = 0$ for $v_l = 0$.

Proof of Proposition 7.1 This is an immediate consequence of the Inverse Function Theorem, which, in view of (7.2.12), implies that the mapping

$$R^2 \ni (v_j, v_l) \; \rightarrow \; z_i(T) \in R^2,$$

defined on some (open) neighborhood of the origin, has the inverse mapping, defined on some (open) neighborhood of $z_i^*(T)$, that is, (7.2.11) holds.

The same argument implies a similar result for the motion of the center of mass $z_c(t)$.

Proposition 7.2 (Local Controllability of z_c) *Let Assumptions 7.1 and 7.2 hold. Assume that in (7.2.1)–(7.2.4) only two distinct control forces are active, say, defined by v_j and v_l (or v_j and w_l, or w_j and w_l), while all other v_k's and w_m's are zeros. Assume further that v_j and v_l are independent of time. Then, if for some $i \in \{1, \ldots, n\}$ there exists a $T > 0$ such that the matrix (or the respective matrices associated with the aforementioned alternative pairs $\{v_j, w_l\}$ or $\{w_j, w_l\}$)*

$$\left(\frac{dz_c(T)}{dv_j} \; |_{v_j, v_l = 0}, \; \frac{dz_c(T)}{dv_l} \; |_{v_j, v_l = 0} \right) \tag{7.2.12}$$

is non-degenerate, then the system (7.2.1)–(7.2.4) is locally controllable near its drifting position of the center of mass $z_{cd}(T)$. Namely, there is an $\varepsilon > 0$ such that

$$B_\varepsilon(z_{cd}(T)) \subset \{z_c(T) \mid v_j, v_l \in R, \; v_k = 0 \text{ for } k = 1, \ldots, n-2, k \neq j, l; \; w_m = 0, m = 1, \ldots, n-1\}.$$

In particular, for the initial equilibrium position (7.1.3) we have the local controllability with respect to z_c near equilibrium at time T.

7.2.2 Main Results in Terms of Projections of Swimmers' Forces on the Fluid Velocity Space

To formulate our further results, we will need to introduce some notations.

Recall first that the unique solution to (7.2.1)–(7.2.4), see Theorems 5.1 and Corollary 5.1, admits the following implicit Fourier series representation:

$$u(t, x) = \sum_{k=1}^{\infty} e^{-\lambda_k t} \left(\int_\Omega u_0' \omega_k dq \right) \omega_k(x)$$

$$+ \sum_{k=1}^{\infty} \int_{0}^{t} e^{-\lambda_k(t-\tau)} \left(\int_{\Omega} F' \omega_k dq \right) d\tau \omega_k(x). \qquad (7.2.13)$$

Here $'$ stands for transposition, while the 2D vector functions $\omega_k, k = 1, \ldots$ and the real numbers $-\lambda_k, k = 1, \ldots$ denote, respectively, the orthonormalized in $(L^2(\Omega))^2$ eigenfunctions and eigenvalues of the following spectral problem associated with (7.2.1)–(7.2.4):

$$\begin{cases} \nu \Delta \omega_k - \nabla p_k = -\lambda_k \omega_k & \text{in } \Omega, \\ \text{div } \omega_k = 0 & \text{in } \Omega, \\ \omega_k = 0 & \text{in } \partial\Omega, \end{cases} \qquad (7.2.14)$$

where the eigenvalues $-\lambda_k$'s are all negative, of finite multiplicity and tend to $-\infty$ as $k \to \infty$ [36, p. 45]. The series in (7.2.13) converges in V uniformly for $t \in (0, T)$, while the series obtained from it by differentiation once with respect to t and twice with respect to the spatial variables converge in $(L^2(Q_T))^2$, (e.g., [36], [56], [44, Mikhailov, p. 377]). The functions $\{\omega_k\}_{k=1}^{\infty}$ also form a basis in $H \cap V$.

Denote the sum of all terms in F in (7.2.1)–(7.2.4), generated at the initial moment $t = 0$ by the unit control input $v_j = 1$ (alternatively $w_k = 1$, see Propositions 7.1 and 7.2), by $F_j(x) = F_j(0, x)$ and its projection of on the *divergence-free* space H by $F_{j,1}(x)$.

Thus,

$$F_{j,1}(x) = \sum_{k=1}^{\infty} \int_{\Omega} F'_j \omega_k \, dq \, \omega_k(x). \qquad (7.2.15)$$

Recall now that the space $(L^2(\Omega))^2$ is the direct sum of the spaces H and $G(\Omega)$. Decompose $F_j(x)$, respectively:

$$F_j(x) = F_{j,1}(x) + F_{j,2}(x), \quad F_{j,1} \in H, \quad F_{j,2} \in G(\Omega).$$

Since (e.g., [36, p. 28]; [56, p. 15]) :

$$H = \{u \in (L^2(\Omega))^2, \ \text{div } u = 0, \ \gamma_\nu u \mid_{\partial\Omega} = 0\}, \qquad (7.2.16)$$

$$G(\Omega) = \{u \in (L^2(\Omega))^2, \ u = \nabla p, \ p \in H^1(\Omega)\}, \qquad (7.2.17)$$

where ν is the unit vector normal to the boundary $\partial\Omega$ (pointing outward) and $\gamma_\nu u \mid_{\partial\Omega}$ is the restriction of $u \cdot \nu$ to $\partial\Omega$, we have

$$F_{j,2}(x) = \nabla p_j(x)$$

for some functions w_j. Thus,

$$F_{j,1}(x) = F_j(x) - \nabla p_j(x). \tag{7.2.18}$$

Furthermore, p_j solves the following generalized Neumann boundary problem:

$$\Delta w_j = \text{div } F_j \quad \text{in } \Omega, \quad \frac{\partial p_j}{\partial \nu}\bigg|_{\partial\Omega} = 0. \tag{7.2.19}$$

Indeed, the Eq. (7.2.19), e.g., can be obtained by applying divergence to (7.2.18) and recalling that $F_{j,1} \in H$, i.e., div $F_j, 1 = 0$. In turn, the boundary condition in (7.2.19) follows from (7.2.18) by recalling that, due to (7.2.16), $\gamma_\nu F_{j,1}|_{\partial\Omega} = 0$ and that the characteristic functions of supports $S(z_k(0))$'s for $F_j(x)$ vanish outside of these sets.

Assumption 7.3 *Let the $[2 \times 2]$-matrix*

$$\left(\int_{S(z_i(0))} F_{j,1}(x)dx, \quad \int_{S(z_i(0))} F_{l,1}(x)dx \right) \tag{7.2.20}$$

be non-degenerate for some $i \in \{1, \ldots, n\}$ and $l, j \in \{1, \ldots, n-1\}$. Alternatively, for other pairs of controls in Propositions 7.1 and 7.2 we require that the respective matrices are non-degenerate.

Remark 7.4 Assumption 7.3 can be easily and explicitly verified in the case when $S(z_i(t))$'s are small narrow rectangles or discs making use of results in Theorems 9.3 and 9.4 in Chap. 9.

Theorem 7.2 *Let $i \in \{1, \ldots, n\}$, $l, j \in \{1, \ldots, n-2\}$ and Assumptions 7.1– 7.3 and of Proposition 7.1 hold. Then there exists a $T^* > 0$ such that the matrix (7.2.12) is non-degenerate for any $T \in (0, T^*]$ and Proposition 7.1 holds. Namely, we have the local controllability of system (7.2.1)–(7.2.4) near its drifting position $z_i^*(T), T \in (0, T^*]$. In particular, for the equilibrium position (7.1.3)–(7.1.4) condition (7.2.20) implies the local controllability with respect to z_i near equilibrium at time T. The above holds for alternative pairs of controls in Propositions 7.1 and 7.2 as well.*

The argument of Theorem 7.2 establishes that

$$\frac{dz_i(t)}{dv_j}\bigg|_{v_j, v_l=0} = \frac{t^2}{2\text{meas}\{S\}} \int_{S(z_i(0))} F_{j,1}(x)dx + t^2 O(t), \tag{7.2.21}$$

(or similar expression for $\frac{dz_i(t)}{dw_j}\big|_{v_k's, w_m's=0}$), which allows us to apply (7.2.20) to ensure that (7.2.12) in Proposition 7.1 is non-degenerate.

Due to (7.2.21), at no extra cost, Theorem 7.2 implies the respective statement for the center of mass $z_c(t)$.

Theorem 7.3 *Let Assumptions 7.1–7.3 hold. Let* $l, j \in \{1, \ldots, n - 2\}$ *and the matrix*

$$\sum_{i=1}^{n} \left(\int_{S(z_i(0))} F_{j,1}(x)dx, \quad \int_{S(z_i(0))} F_{l,1}(x)dx \right)$$

be non-degenerate. Then there exists a $T^* > 0$ *such that for any* $T \in (0, T^*]$ *Proposition 7.2 holds. Namely, we have the local controllability of system (7.2.1)– (7.2.4) with respect to the position of center of mass* $z_c(T)$ *near its drifting position* $z_c^*(T), T \in (0, T^*]$. *In particular, for the equilibrium position (7.1.3)–(7.1.4) we have the local controllability with respect to* z_c *near equilibrium at time* T. *The above holds for alternative pairs of controls in Propositions 7.1–7.2 as well.*

7.3   Preliminary Results

Our plan to prove Theorem 7.2 is as follows:

- In Sect. 7.3 we intend to differentiate the implicit solution formula (7.2.13) with respect to v_j's and w_k's, satisfying (7.2.7), to derive the formulas for the respective derivatives in Propositions 7.1 and 7.2, namely:

$$\frac{dz_i(T)}{dv_j}, \quad \frac{dz_i(T)}{dw_k}, \quad i = 1, \ldots, n, \, j = 1, \ldots, n - 2, \, k = 1, \ldots, n - 1.$$

- In Sect. 7.4 these results will be presented in the form a suitable Volterra equations for $\frac{dz_i(T)}{dv_j}$'s and $\frac{dz_i(T)}{dw_k}$'s, with the goal to evaluate their asymptotic behavior as $T \to 0$.
- Making use of all of the above, to obtain the qualitative estimates for the terms in (7.2.20), we complete the proof of Theorem 7.2 in Sect. 7.6.

7.3.1   Implicit Solution Formula

Let us rewrite (7.2.13) in the following form:

$$u(t, x) = \sum_{k=1}^{\infty} e^{-\lambda_k t} \left(\int_{\Omega} u_0^T \omega_k dx \right) \omega_k(x) + \sum_{i=1}^{n-2} v_i(t) Q_i(t, x) + \sum_{i=1}^{n-1} w_i(t) R_i(t, x),$$

(7.3.1)

where $Q_i(t, x)$'s and $P_k(t, x)$'s are calculated along the expressions in (7.2.3) and (7.2.4).

7.3.2 Differentiation with Respect to v_j's and w_k's

Below we will focus on the derivatives with respect to v_j's (calculations for w_k's are similar). Our calculations are formal in part in this section with further justification discussed in Sect. 7.5.

We assume that v_j, $j = 1, \ldots, n - 2$ and w_k, $k = 1, \ldots, n - 1$ are independent variables in R (see (7.2.7) in Assumption 7.2).

Derivatives $\frac{dz_i(t)}{dv_j}$**'s** First, we will write the equations for z_i's in (7.2.3) in their integral form:

$$z_i(t) = z_{i,0} + \frac{1}{\text{meas}\{S\}} \int_0^t \int_{S(z_i(\tau))} u(\tau, x) dx d\tau.$$

$$= z_{i,0} + \frac{1}{\text{meas}\{S\}} \int_0^t \int_{z_{i,1}(\tau)-r}^{z_{i,1}(\tau)+r} \int_{z_{i,2}(\tau)+\alpha_i(\tau,x_1-z_{i,1}(\tau))}^{z_{i,2}(\tau)+\beta_i(\tau,x_1-z_{i,1}(\tau))} u(\tau, x) dx_2 dx_1 d\tau.$$

(7.3.2)

Differentiating (7.3.2) with respect to v_j (that is, taking a partial derivative), we obtain

$$\frac{dz_i(t)}{dv_j} = \frac{1}{\text{meas}\{S\}} \int_0^t \int_{z_{i,1}(\tau)-r}^{z_{i,1}(\tau)+r} (u(\tau, x_1, z_{i,2}(\tau) + \beta_i(\tau, x_1 - z_{i,1}(\tau)))$$

$$\times \left(\frac{dz_{i,2}(\tau)}{dv_j} - \beta_{ix_1}(\tau, x_1 - z_{i,1}(\tau)) \frac{dz_{i,1}(\tau)}{dv_j} \right) - u(\tau, x_1, z_{i,2}(\tau))$$

$$+ \alpha_i(\tau, x_1 - z_{i,1}(\tau))) \left(\frac{dz_{i,2}(\tau)}{dv_j} - \alpha_{ix_1}(\tau, x_1 - z_{i,1}(\tau)) \frac{dz_{i,1}(\tau)}{dv_j} \right) dx_1 d\tau$$

$$+ \frac{1}{\text{meas}\{S\}} \int_0^t \frac{dz_{i,1}(\tau)}{dv_j} \int_{z_{i,2}(\tau)+\alpha_i(\tau,r)}^{z_{i,2}(\tau)+\beta_i(\tau,r)} u(\tau, z_{i,1}(\tau) + r, x_2) dx_2 d\tau$$

$$- \frac{1}{\text{meas}\{S\}} \int_0^t \frac{dz_{i,1}(\tau)}{dv_j} \int_{z_{i,2}(\tau)+\alpha_i(\tau,-r)}^{z_{i,2}(\tau)+\beta_i(\tau,-r)} u(\tau, z_{i,1}(\tau) - r, x_2) dx_2 d\tau$$

$$+ \frac{1}{\text{meas}\{S\}} \int_0^t \int_{S(z_i(\tau))} \frac{du(\tau, x)}{dv_j} dx d\tau. \tag{7.3.3}$$

(Partial) Derivatives $\frac{du}{dv_j}$**'s** Making use of (7.3.1), we obtain

$$\frac{du}{dv_j} |_{v'_k s, w'_m s = 0} = \frac{d}{dv_j} \left(\sum_{i=1}^{n-2} v_i Q_i(t, x) + \sum_{i=1}^{n-1} w_i R_i(t, x) \right) |_{v'_k s, w'_m s = 0}$$

$$= Q_j(t, x) + \left(\sum_{i=1}^{n-2} v_i \frac{d}{dv_j} Q_i(t, x) + \sum_{i=1}^{n-1} w_i \frac{d}{dv_j} R_i(t, x) \right) |_{v'_k s, w'_m s = 0} = Q_j(t, x).$$

$$\tag{7.3.4}$$

Here, the subscript $v'_k s, w'_m s = 0$ indicates that we calculate the partial derivative with respect to v_j at $v_j = 0$, for $v_k = 0, k = 1, \ldots, n - 2, k \neq j, w_m = 0, m = 1, \ldots, n - 1$.

7.4   Volterra Equations for $\frac{dz_i(\tau)}{dv_j}$'s

Below we will deal only with the terms $\frac{dz_i(t)}{dv_j} |_{v_j, v_l=0}$. Therefore, to simplify further notations, we will *omit the subscript* $|_{v'_k s, w'_m s = 0}$ *from now on.*

Let us rewrite (7.3.3) and (7.3.4) as the following Volterra equation:

$$\frac{dz_i(t)}{dv_j} + \int_0^t \mathbf{B}_j(t, s) \frac{dz_i(s)}{dv_j} ds = \frac{1}{\text{meas}\{S\}} \int_0^t \int_{S(z_i(\tau))} Q_j dx d\tau, \quad t \in [0, T],$$

$$\tag{7.4.1}$$

where the matrix $\mathbf{B}_j(t, s)$ is defined by (7.3.3) and (7.3.4).

Now, due to (7.2.13), (7.2.3), Theorem 5.1, (7.2.6) and (7.2.15):

$$\frac{1}{\text{meas}\{S\}} \int_0^t \int_{S(z_i(\tau))} Q_j dx d\tau$$

$$= \frac{1}{\text{meas}\{S\}} \sum_{k=1}^{\infty} \int_0^t \int_{S(z_i(\tau))} \int_0^{\tau} e^{-\lambda_k(\tau-s)} \left(\int_{\Omega} F_j(s, q)' \omega_k dq \right) \omega_k(x) ds dx d\tau$$

$$= \frac{t^2}{2\text{meas}\,\{S\}} \sum_{k=1}^{\infty} \int_{S(z_i(0))} \left(\int_{\Omega} F_j(0, q)' \omega_k dq \right) \omega_k(x) dx + t^2 O(t)$$

$$= \frac{t^2}{2\text{meas}\,\{S\}} \int_{S(z_i(0))} F_{j,1}(x) dx + t^2 O(t), \tag{7.4.2}$$

where $O(t)$ stands for an expression that tends to zero as $t \to 0+$ and

$$F_j(t, x) = \Big[\xi_j(t, x) A(z_j(t) - z_{j+1}(t))$$

$$- \xi_{j+2}(t, x) \frac{\|z_j(t) - z_{j+1}(t)\|^2}{\|z_{j+2}(t) - z_{j+1}(t)\|^2} A(z_{j+2}(t) - z_{j+1}(t)) \Big]$$

$$+ \xi_{j+1}(t, x) \left[A(z_{j+1}(t) - z_j(t)) - \frac{\|z_j(t) - z_{j+1}(t)\|^2}{\|z_{j+2}(t) - z_{j+1}(t)\|^2} A(z_{j+1}(t) - z_{j+2}(t)) \right].$$

$$\tag{7.4.3}$$

Let us elaborate on the derivation of (7.4.2).

Step 1 Note that $Q_j(t, x)$ solves the following initial- and boundary value problem:

$$\begin{cases} Q_{jt} = v \wedge Q_j + F_j(t, x) - \nabla p_Q & \text{in } Q_T = (0, T) \times \Omega, \\ \text{div}\, Q_j = 0 & \text{in } Q_T, \\ Q_j = 0 & \text{in } \Sigma_T = (0, T) \times \partial\Omega, \\ Q_j(0, \cdot) = 0 & \text{in } \Omega. \end{cases} \tag{7.4.4}$$

Let, in turn, Q_j^0 denote the solution to

$$\begin{cases} Qj_{jt}^0 = v \Delta Q^0 + F_j(0, x) - \nabla p_Q^0 & \text{in } Q_T, \\ \text{div}\, Q^0 = 0 & \text{in } Q_T, \\ Q_j^0 = 0 & \text{in } \Sigma_T = (0, T) \times \partial\Omega, \\ Q_j^0(0, \cdot) = 0 & \text{in } \Omega. \end{cases} \tag{7.4.5}$$

Then, formula (7.2.13) applied to $(Q - Q^0)$, along with Bessel's inequality, gives (compare to (5.5.15)), the following estimate:

$$\|Q - Q^0)\|_{C([0,t]:H)} \leq \sqrt{t} \|F_j - F_j(0, \cdot)\|_{(L^2(Q_t))^2}$$

$$\leq t \sqrt{\text{meas}\,\{S\}} \|F_j - F_j(0, \cdot)\|_{C([0,T)];(L^2(\Omega))^2)}. \tag{7.4.6}$$

Step 2 Next,

$$F_j(t, x) - F_j(0, x)$$

$$= \begin{cases} F_j(t, x) - F_j(0, x), & \text{on their common support,} \\ +\text{the remaining bounded terms, supported within the sets} \\ S(z_j(t))\Delta S(z_j(0)) \bigcup S(z_{j+1}(t))\Delta S(z_{j+1}(0)) \bigcup S(z_{j+2}(t))\Delta S(z_{j+2}(0)), \end{cases}$$
(7.4.7)

where $A\Delta B$ stands for symmetric set difference.

Due to continuity of $z_i(t)$'s in Theorem 5.1 and (7.2.6) in Assumption 7.1, (7.4.7) yields

$$\|F_j - F_j(0, \cdot)\|_{(C([0,T)];(L^2(\Omega))^2)} = O(t).$$
(7.4.8)

Combining (7.4.6)–(7.4.8), we obtain that (with $F_j(x) = F_j(0, x)$):

$$\frac{1}{\text{meas}\{S\}} \int_0^t \int_{S(z_i(\tau))} Q_j(\tau, x)dxd\tau$$

$$= \frac{1}{\text{meas}\{S\}} \sum_{k=1}^{\infty} \int_0^t \int_{S(z_i(\tau))} \int_0^{\tau} e^{-\lambda_k(\tau-s)} \left(\int_{\Omega} F_j(s, q)'\omega_k dq \right) \omega_k(x)dsdxd\tau$$

$$= \frac{1}{\text{meas}\{S\}} \int_0^t \int_{S(z_i(\tau))} Q_j^0(t, x)dxd\tau + t^2 O(t)$$

$$= \frac{1}{\text{meas}\{S\}} \sum_{k=1}^{\infty} \int_0^t \int_0^{\tau} e^{-\lambda_k(\tau-s)}ds \int_{S(z_i(\tau))} \left(\int_{\Omega} F_j(0, q)'\omega_k dq \right) \omega_k(x)dxd\tau + t^2 O(t)$$

$$= \frac{1}{\text{meas}\{S\}} \int_0^t \int_{S(z_i(0))} Q_j^0(t, x)dxd\tau + t^2 O(t)$$

$$= \frac{1}{\text{meas}\{S\}} \sum_{k=1}^{\infty} \int_0^t \int_0^{\tau} e^{-\lambda_k(\tau-s)}ds \int_{S(z_i(0))} \left(\int_{\Omega} F_j(0, q)'\omega_k dq \right) \omega_k(x)dxd\tau + t^2 O(t)$$

$$= \frac{1}{\text{meas}\{S\}} \sum_{k=1}^{\infty} \int_0^t \int_0^{\tau} ds \int_{S(z_i(0))} \left(\int_{\Omega} F_j(q)'\omega_k dq \right) \omega_k(x)dxd\tau + t^2 O(t),$$
(7.4.9)

from which (7.4.2) follows.

7.5   Auxiliary Estimates

It is well known that, as a special version of Fredholm equation, (7.4.1) admits a unique solution in $(L^2(0, T))^2$. This will allow us to prove, making use of the classical methods, that dz_i/dv_j indeed exists in $(L^2(0, T))^2$, and that all the above calculations leading to (7.4.1) are valid.

To this end, one needs to write the Volterra equation for the expression $\Delta z_i/\Delta v_j$ and then pass to the limit as Δv_j tends to zero to obtain (7.4.1), we refer to Chap. 8 for this technique in more detail in the framework of Navier–Stokes equations.

Furthermore, there exists a ("small") $T_1 > 0$ such that for any $t \in (0, T_1], i = 1, \ldots, n, j = 1, \ldots, n - 2$:

$$\| \frac{dz_i}{dv_j} \|_{(L^2(0,t))^2} \leq C \| \, \Xi_j \, \|_{(L^2(0,t))^2}, \quad \Xi_j(t) = \frac{1}{\text{meas} \{S\}} \int_0^t \int_{S(z_i(\tau))} Q_j dx d\tau,$$

(7.5.1)

where $C > 0$ is a (generic) positive constant independent of $t \in [0, T_1]$.

Moreover, since in (7.4.1) the integral term with $(L^2(0, t))^2$-derivative dz_i/dv_j and the right-hand sides are actually continuous functions, we have

$$\frac{dz_i}{dv_j} \in C([0, t]; R^2)$$

and, similar to (7.5.1), making use of (7.4.2), we can derive that there exists a $T_2 \in [0, T_1]$ such that for any $t \in (0, T_2]$:

$$\| \frac{dz_i}{dv_j} \|_{C([0,t]; R^2)} \leq Lt^2$$

(7.5.2)

for some constant $L > 0$ as $t \to 0+$.

7.6   Proof of Theorem 7.2

From (7.4.1), making use of the above, we can derive that

$$\frac{dz_i(t)}{dv_j} = \frac{t^2}{2\text{meas} \{S\}} \int_{S(z_i(0))} F_j(x)dx + t^2 O(t)$$

(7.6.1)

for $j = 1, \ldots, n - 2$ as $t \to 0$.

Thus, the matrix

$$\left(\frac{dz_i(t)}{dv_j}, \frac{dz_i(t)}{dv_l} \right)$$

is not degenerate for sufficiently small t. We can now select any such "small" number t as a $T > 0$ in Proposition 7.1 and obtain the statement of Theorem 7.2 from this proposition. This ends the proof of Theorem 7.2.

7.7 Proof of Theorem 7.1

We have form (7.2.13):

$$\frac{1}{\text{meas}\,\{S\}} \int\limits_{S(z_i(0))} u(t,x)dx = \sum_{k=1}^{\infty} \frac{1}{\text{meas}\,\{S\}} \int\limits_{S(z_i(0))} e^{-\lambda_k t} \left(\int\limits_{\Omega} u_0' \omega_k ds \right) \omega_k(x)dx$$

$$+ \sum_{k=1}^{\infty} \frac{1}{\text{meas}\,\{S\}} \int\limits_{S(z_i(0))} \int\limits_{0}^{t} e^{-\lambda_k(t-\tau)} \left(\int\limits_{\Omega} (P_H F)' \omega_k ds d\tau \right) \omega_k(x)dx. \qquad (7.7.1)$$

7.7.1 Step 1

We will deal first with the first that the first term in (7.7.1):

$$\frac{1}{\text{meas}\,\{S\}} \int\limits_{S(z_i(t))} \left(\sum_{k=1}^{\infty} e^{-\lambda_k t} \left(\int\limits_{\Omega} u_0' \omega_k ds \right) \omega_k(x) \right) dx$$

$$= \frac{1}{\text{meas}\,\{S\}} \int\limits_{S(z_i(0))} u_0 dx + p(t), \qquad (7.7.2)$$

where

$$\| p(t) \|_{R^2} = \left\| \frac{1}{\text{meas}\,\{S\}} \int\limits_{S(z_i(t))} \left(\sum_{k=1}^{\infty} e^{-\lambda_k t} \left(\int\limits_{\Omega} u_0' \omega_k ds \right) \omega_k(x) \right) dx$$

$$- \frac{1}{\text{meas}\,\{S\}} \int\limits_{S(z_i(0))} u_0 dx \,\|_{R^2}$$

$$\leq \frac{1}{\text{meas}\,\{S\}} \| \int\limits_{S(z_i(0)) \bigcap S(z_i(t))} \sum_{k=1}^{\infty} \left(e^{-\lambda_k t} - 1\right) \int\limits_{\Omega} u_0' \omega_k ds \omega_k(x) \, dx \,\|_{R^2}$$

$$+ \frac{1}{\text{meas}\,\{S\}} \| \int\limits_{S(z_i(0)) \backslash S(z_i(t))} \sum_{k=1}^{\infty} \int\limits_{\Omega} u_0' \omega_k ds \omega_k(x) \, dx \,\|_{R^2}$$

$$+ \frac{1}{\text{meas}\,\{S\}} \| \int\limits_{S(z_i(t)) \backslash S(z_i(0))} \sum_{k=1}^{\infty} e^{-\lambda_k t} \int\limits_{\Omega} u_0' \omega_k ds \omega_k(x) \, dx \,\|_{R^2}$$

$$\leq \frac{\mu(t, u_0)}{\text{meas}\,\{S\}} \text{meas}^{1/2}\,\{S(z_i(0)) \bigcap S(z_i(t))\}$$

$$+ \frac{2\max^{1/2}\{\text{meas}\,\{S(z_i(0))\backslash S(z_i(t)), \ \text{meas}\, S(z_i(t))\backslash S(z_i(0))\}}{\text{meas}\,\{S\}} \| u_0 \|_{(L^2(\Omega))^2},$$

$$\leq \frac{\mu(t, u_0)}{\text{meas}^{1/2}\,\{S\}} + \frac{2\hat{s}^{1/2}(t) \| u_0 \|_{(L^2(\Omega))^2}}{\text{meas}\,\{S\}}, \qquad (7.7.3)$$

where $\mu(t, u_0) \to 0$ as $t \to 0+, 0 \leq \mu(t, u_0) \leq \| u_0 \|_{(L^2(\Omega))^2}$.

7.7.2 Step 2

All the terms in (7.7.1) associated with the forcing term F from (7.2.3) and (7.2.4) admit the following representation:

$$\sum_{k=1}^{\infty} \int\limits_0^t e^{-\lambda_k(t-\tau)} \left(\int\limits_{\Omega} \xi_i(\tau, s) f'(\tau) \omega_k ds d\tau \right) \omega_k(x) \qquad (7.7.4)$$

for $i \in \{1, \ldots, n\}$ and some 2D vector-function $f(t)$.

We need to evaluate, e.g., the following expression (compare to (7.2.8)):

$$\| \frac{1}{\text{meas}\,\{S\}} \int\limits_{S(z_i(t))} \left(\sum_{k=1}^{\infty} \int\limits_0^t e^{-\lambda_k(t-\tau)} \left(\int\limits_{\Omega} \xi_i(\tau, s) f'(\tau) \omega_k ds d\tau \right) \omega_k(x) \right) dx$$

$$- \frac{t}{\text{meas}\{S\}} \int\limits_{S(z_i(0))} \sum_{k=1}^{\infty} \left(\int\limits_{\Omega} \xi_i(\tau, s) f'(0) \omega_k ds \right) \omega_k(x) dx \, \|_{R^2}$$

$$\leq \frac{1}{\text{meas}\{S\}} \| \int\limits_{S(z_i(0)) \bigcap S(z_i(t))} (\sum_{k=1}^{\infty} \int_0^t e^{-\lambda_k(t-\tau)}$$

$$\times (\int\limits_{\Omega} [\xi_i(\tau, s) w(\tau) - \xi_i(0, s) f(0)]' \omega_k ds d\tau) \omega_k(x)) dx \, \|_{R^2}$$

$$+ \frac{1}{\text{meas}\{S\}} \| \int\limits_{S(z_i(0)) \bigcap S(z_i(t))} (\sum_{k=1}^{\infty} \int_0^t [e^{-\lambda_k(t-\tau)} - 1]$$

$$\times (\int\limits_{\Omega} \xi_i(0, s) f'(0) \omega_k ds d\tau) \omega_k(x)) dx \, \|_{R^2}$$

$$+ \frac{1}{\text{meas}\{S\}}$$

$$\times \| \int\limits_{S(z_i(t)) \backslash S(z_i(0))} \left(\sum_{k=1}^{\infty} \int_0^t e^{-\lambda_k(t-\tau)} \left(\int\limits_{\Omega} \xi_i(\tau, s) f'(\tau) \omega_k ds d\tau \right) \omega_k(x) \right) dx \, \|_{R^2}$$

$$+ \frac{t}{\text{meas}\{S\}} \| \int\limits_{S(z_i(0)) \backslash S(z_i(t))} \left(\sum_{k=1}^{\infty} \left(\int\limits_{\Omega} \xi_i(\tau, s) f'(0) \omega_k ds \right) \omega_k(x) \right) dx \, \|_{R^2}$$

$$\leq \frac{t^{1/2} \text{meas}^{1/2} \{S(z_i(0)) \bigcap S(z_i(t))\}}{\text{meas}\{S\}} \| \xi_i(\cdot, \cdot) f(\cdot) - \xi_i(0, \cdot) f(0) \|_{(L^2(Q_t))^2}$$

$$+ \frac{1}{\text{meas}^{1/2}\{S\}} \left(\sum_{k=1}^{\infty} t^2 \left[\frac{1 - e^{-\lambda_k t}}{\lambda_k t} - 1 \right]^2 \left(\int\limits_{\Omega} \xi_i(0, s) f'(0) \omega_k ds \right)^2 \right)^{1/2}$$

$$+ \frac{t^{1/2} \text{meas}^{1/2} \{S(z_i(t)) \backslash S(z_i(0))\}}{\text{meas}\{S\}} \| \xi_i(\cdot, \cdot) f(\cdot) \|_{(L^2(Q_t))^2}$$

$$+ t \, \frac{\text{meas}^{1/2} \{S(z_i(0)) \backslash S(z_i(t))\}}{\text{meas} \{S\}} \, \| \, \xi_i(\cdot, 0) f(0) \, \|_{(L^2(\Omega))^2}$$

$$\leq t \, \frac{\text{meas} \{S(z_i(0)) \bigcap S(z_i(t))\}}{\text{meas} \{S\}} \, \left(\| \, f(\cdot) - f(0) \, \|_{(C[0,t])^2} \right) \, + \, t \frac{1}{\text{meas}^{1/2} \{S\}} \gamma(t)$$

$$+ t \, \frac{\text{meas}^{1/2} \{S(z_i(t)) \backslash S(z_i(0))\}}{\text{meas} \{S\}} \text{meas}^{1/2} \{S(z_i(0)) \bigcap S(z_i(t))\} \, \| \, f(\cdot) \, \|_{(C[0,t])^2}$$

$$+ t \, \frac{\text{meas}^{1/2} \{S(z_i(0)) \backslash S(z_i(t))\}}{\text{meas} \{S\}} \text{meas}^{1/2} \{S(z_i(0)) \bigcap S(z_i(t))\} \, \| \, f(0) \, \|_{R^2}$$

$$+ t \, \frac{\text{meas}^{1/2} \{S(z_i(t)) \backslash S(z_i(0))\}}{\text{meas}^{1/2} \{S\}} \, \| \, f(\cdot) \, \|_{(C[0,t])^2}$$

$$+ t \, \frac{\text{meas}^{1/2} \{S(z_i(0)) \backslash S(z_i(t))\}}{\text{meas}^{1/2} \{S\}} \, \| \, f(0) \, \|_{R^2} \leq C \, \| \, f(\cdot) \, \|_{(C[0,t])^2} \, o(t) \quad \text{as } t \to 0+,$$

$$(7.7.5)$$

where we used (7.2.6) in Assumption 7.1, estimate

$$| \, \frac{1 - e^{-s}}{s} - 1 \, | < 1, \quad s > 0$$

and

$$\gamma(t) = \left(\sum_{k=1}^{\infty} \left[\frac{1 - e^{-\lambda_k t}}{\lambda_k t} - 1 \right]^2 \left(\int_{\Omega} \xi_i(0, s) f'(0) \omega_k ds \right)^2 \right)^{1/2} \to 0+ \text{ as } t \to 0+.$$

The estimates (7.7.3) and (7.7.5) yield the statement of Theorem 7.1.

Chapter 8
Local Controllability of 2D and 3D Swimmers: The Case of Navier–Stokes Equations

In this chapter we will extend the method of Chap. 7 to the swimmers in a fluid described by the 2D and 3D Navier–Stokes equations. Since the implicit generalized Fourier series representation for solutions, employed in Chap. 7 for the linear non-stationary Stokes equations, is not an option for the nonlinear Navier–Stokes equations, we will elaborate on a more general scheme of proofs, already mentioned in Sect. 7.5.

8.1 Problem Setting

In this chapter we consider the 2D and 3D models from Chap. 6:

$$
\begin{cases}
u_t + (u \cdot \nabla)u = \nu \Delta u + F - \nabla p & \text{in } Q_T, \\
\operatorname{div} u = 0 & \text{in } Q_T, \\
u = 0 & \text{in } \Sigma_T = (0, T) \times \partial\Omega, \\
u(0, \cdot) = u_0 \in V & \text{in } \Omega \subset R^K, K = 2, 3 \\
\frac{dz_i}{dt} = \frac{1}{\operatorname{meas}(S)} \int_{S(z_i(t))} u(t, x)dx, \ z_i(0) = z_{i,0}, \ i = 1, \ldots, n,
\end{cases}
$$

$$(8.1.1)$$

where $x = (x_1, \ldots, x_K)$, $u(t, x) = (u_1(t, x), \ldots, u_K(t, x))$,

$$z(t) = (z_1(t), \ldots, z_n(t)) \in [R^K]^n, \quad v(t) = (v_1(t), \ldots, v_{n-2}(t)) \in R^{n-2},$$

$$w(t) = (w_1(t), \ldots, w_{n-1}(t)) \in R^{n-1},$$

a set of geometric controls, as in (5.2.2),

$$S_i(0, t), \quad t \in [0, T], \quad i = 1, \ldots, n, \tag{8.1.2}$$

A. Khapalov, *Bio-Mimetic Swimmers in Incompressible Fluids*, Lecture Notes in Mathematical Fluid Mechanics, https://doi.org/10.1007/978-3-030-85285-6_8

is prescribed, and F can be of any configuration of the following swimmers' internal *control* rotational and elastic forces (no Hooke's uncontrolled forces, see Remark 7.1), introduced in Chap. 2, namely

- for $K = 2$:

$$F_{rot2D}(t, x) = \sum_{i=2}^{n-1} v_{i-1}(t) \left[\xi_{i-1}(t, x) A(z_{i-1}(t) - z_i(t)) \right.$$

$$\left. - \xi_{i+1}(t, x) \frac{\|z_{i-1}(t) - z_i(t)\|^2}{\|z_{i+1}(t) - z_i(t)\|^2} A(z_{i+1}(t) - z_i(t)) \right]$$

$$+ \sum_{i=2}^{n-1} \xi_i(t, x) v_{i-1}(t) \left[A(z_i(t) - z_{i-1}(t)) - \frac{\|z_{i-1}(t) - z_i(t)\|^2}{\|z_{i+1}(t) - z_i(t)\|^2} A(z_i(t) - z_{i+1}(t)) \right],$$
(8.1.3)

- for $K = 3$:

$$F_{rot3D}(t, x) = \sum_{i=2}^{N-1} v_{i-1}(t) \left[\xi_{i-1}(t, x) \ A_i(t)(z_{i-1}(t) - z_i(t)) \right.$$

$$\left. - \xi_{i+1}(t, x) \frac{|z_{i-1}(t) - z_i(t)|^2}{|z_{i+1}(t) - z_i(t)|^2} \ B_i(t)(z_{i+1}(t) - z_i(t)) \right]$$

$$+ \sum_{i=2}^{N-1} \xi_i(t, x) v_{i-1}(t) \left[A_i(t)(z_i(t) - z_{i-1}(t)) - \frac{|z_{i-1}(t) - z_i(t)|^2}{|z_{i+1}(t) - z_i(t)|^2} \ B_i(t)(z_i(t) - z_{i+1}(t)) \right],$$
(8.1.4)

- and, respectively, for $K = 2, 3$:

$$F_{cef2Dor3D}(t, x) = \sum_{i=2}^{n} [\xi_{i-1}(t, x) w_{i-1}(z_i(t) - z_{i-1}(t)) + \xi_i(t, x) w_{i-1}(z_{i-1}(t) - z_i(t))].$$
(8.1.5)

To simplify further notations, lets us introduce the following extended vector for all swimmer's control forces:

$$v = (v_1, \ldots, v_{n-2}, w_1, \ldots, w_{n-1}) \overset{\Delta}{=} (v_1, \ldots, v_{2n-3}).$$
(8.1.6)

Respectively,

$$F = \sum_{i=1}^{2n-3} v_i F_i.$$
(8.1.7)

Equation (8.1.1) in the above is understood in the sense of the following identity [36, 37]:

$$\int_{\Omega} u(x,t)\phi(t)dx - \int_{\Omega} u(x,0)\phi(t)dx - \int_0^t \int_{\Omega} u \cdot \phi_t dx dt =$$

$$= -\int_0^t v[u(\tau,\cdot),\phi(\tau,\cdot)]dt + \int_0^t \int_{\Omega} \left(-(u \cdot \nabla)u + P_H F_j(\tau,x)\right)\phi dx d\tau, \quad t \in [0,T],$$

(8.1.8)

where

•

$$[\varphi,\psi] = \sum_{i,j=1}^K \int_{\Omega} \varphi_{ix_j} \psi_{ix_j} dx, \quad \varphi = (\varphi_1, \ldots, \varphi_K), \quad \psi = (\varphi_1, \ldots, \varphi_K),$$

• $P_H : [L^2(\Omega)]^K \to H$ denotes the projection operator from $[L^2(\Omega)]^K$ onto H, and
• $\phi \in \Phi = \{\phi \in L^2(0,T;V), \phi_t \in L^2(0,T;H)\}$.

Remark 8.1 In [3] we considered a similar local controllability problem for a different setup of swimmer's models, namely with no geometric controls engaged. In this chapter we intend to follow the guidelines of [3] and of Chap. 7, assuming that the geometric controls of swimmers are active, i.e., the sets $S(z_i(t))$'s can change their spatial orientations relative to their centers $z_i(t)$'s (see Sect. 5.2).

Similar to Assumptions 7.1 and 7.2 in Chap. 7, in this chapter, we will also impose the following additional assumptions on the sets $S(z_i(t))$'s, besides the assumptions in Sect. 5.2.

Assumption 8.1 *We assume, throughout Chap. 8 that, in addition to assumptions in Sect. 5.2:*

•

$$\int_{S_i(0,t)\Delta S_i(0,0)} dx \to 0 \text{ as } t \to 0, \quad i = 1, \ldots, n.$$

(8.1.9)

• *The values of $\|v_i\|_{L^{\infty}(0,T)}$'s and of $\|w_i\|_{L^{\infty}(0,T)}$'s are bounded by some positive number K.*
• *We will use only the constant functions for v_i's:*

$$v_i(t) \equiv L_i, \quad \in \{c \in R| \, |c| \leq \hat{K}\} \, i = 1, \ldots, 2n-3.$$

(8.1.10)

• *T is selected to satisfy Theorems 6.1–6.3 to ensure the well-posedness of system (8.1.1)–(8.1.5) in $[0,T]$, for the given aforementioned constant $\hat{K}$.*

8.2   Main Results

8.2.1   Main Results: Micromotions in 2D and 3D

The following theorem provides the formula for the *initial momentary motions of a swimmer in terms of projections of its initial internal forces on H*.

Theorem 8.1 (Local Motions of Swimmers) *Let $K = 2, 3$ and a set of geometric controls $S_i(0, t)$'s, as in (8.1.2), be given. Under the assumptions of Theorems 6.1– 6.3 and Assumption 8.1, if we set in (8.1.1)–(8.1.5) $u_0 \in H$ for $K = 2$ and $u_0 \in V$ for $K = 3$ (also having in mind (8.1.6) and (8.1.7)):*

$$v_j = ha_j \in R, \; j = 1, \ldots, 2n - 3, \; \sum_{j=1}^{2n-3} a_j^2 = 1,$$

$$\|(v_1, \ldots, v_{2n-3})\|_{R^{2n-3}} = |h| \le 1,$$

$$z_i(t) = z_i(t; h),$$

then

$$z_i(t; h) = z_i(t; 0)$$

$$+ \frac{ht^2}{2\text{meas}(S)} \sum_{j=1}^{2n-3} a_j \int_{S(z_i(\tau;0))} (P_H F_j(0, \cdot))(x)dx + ht^2 \rho(t) + h\zeta(h, t), \; t \in [0, T],$$

$$(8.2.1)$$

where

$$\|\rho\|_{C[0,t]]^K} = O(t), \;\; \|\zeta(h, \cdot)\|_{C[0,t]]^K} = O(h),$$

the expressions $\|\rho\|_{C[0,t]]^K}$ and $\|\zeta(h, \cdot)\|_{C[0,t]]^K}$ are defined by u_0 and $F_j(0, x)$.

Remark 8.2 (Discussion of Theorem 8.1) The formula (8.2.1) states that, momentarily, the swimmer at hand will move approximately in the direction that is *co-linear to the vector of the combined projection force term in (8.2.1)*, namely:

$$\sum_{j=1}^{2n-3} a_j \int_{S(z_i(\tau;0))} (P_H F_j(0, \cdot))(x)dx.$$

This formula is instrumental for the proofs of our local controllability results below.

8.2.2   Main Results: Local Controllability in 2D

The main *controllability* results of this chapter are as follows.

Theorem 8.2 (Local Controllability of z_i's: the 2D Case) *Given $u_0 \in H$, under the assumptions of Theorem 6.1 and Assumption 8.1, let $(u^*, z^* = (z_1^*, \dots, z_n^*))$ be the solution to (8.1.1)–(8.1.5) generated by the zero controls $v_1 = \dots = v_{2n-3} = 0$ on some interval $[0, T^*]$. Let for some $i \in \{1, \dots, n\}$ and $k, l \in \{1, \dots, 2n - 3\}$ the vectors*

$$\int_{S(z_i(0))} P_H F_k(0, \cdot) \, dx, \quad \int_{S(z_i(0))} P_H F_l(0, \cdot) \, dx \quad \text{be linearly independent.}$$

$$(8.2.2)$$

Then there exist $T = T(i, k, \ell) \in (0, T^*]$ and $\varepsilon = \varepsilon(i, k, \ell) > 0$ such that

$$B_\varepsilon(z_i^*(T)) \subseteq \left\{ z_i(T) \mid v_k, v_\ell \in R, \quad \text{while} \quad v_j = 0 \text{ for } j = 1, \dots, 2n - 3, \ j \neq k, \ell \right\}.$$

Remark 8.3 Condition (8.2.2) can be easily and explicitly verified in the case when $S(z_i(t))$'s are small parallelepipeds of certain proportions or balls making use of results in Theorems 10.3–10.5 in Chap. 10.

Discussion of Theorem 8.2 In other words, under the conditions of Theorem 8.2, the point z_i can be steered on some time interval $[0, T]$ from its initial position $z_{i,0} = z_i^*(0)$ to any point within some ball $B_\varepsilon(z_i^*(T))$ of radius $\varepsilon > 0$ with center at the endpoint $z_i^*(T)$ of the "drifting" trajectory $z_i^*(t), t \in [0, T]$ (see Chap. 7).

At no extra cost we will have the following result for the motion of the center of mass of our swimmer.

Theorem 8.3 (Local Swimming Locomotion: 2D Case) *Let in Theorem 8.2 condition (8.2.2) be replaced with the following:*

$$\sum_{i=1}^n \int_{S(z_i(0))} P_H F_k(0, \cdot) \, dx, \quad \sum_{i=1}^n \int_{S(z_i(0))} P_H F_l(0, \cdot) \, dx \quad \text{are linearly independent.}$$

$$(8.2.3)$$

Then the result of Theorem 8.2 holds with respect to the swimmer's center of mass $z_c = \frac{1}{n} \sum_{i=1}^n z_i(t)$, namely:

$$B_\varepsilon(z_{cd}(T)) \subseteq \left\{ z_c(T) \mid v_k, v_\ell \in R, \quad \text{while} \quad v_j = 0 \text{ for } j = 1, \dots, 2n - 3, \ j \neq k, \ell \right\},$$

$$z_{cd} = \frac{1}{n} \sum_{i=1}^n z_i^*(t).$$

8.2.3   Main Results: Local Controllability in 3D

Theorem 8.4 (Local Controllability in 3D) *Given $u_0 \in V$, under the assumptions of Theorems 6.2–6.3 and Assumption 8.1, the results of Theorem 8.2 can be extended to the 3D case, assuming that three controls v_k, v_l, and v_m are active and the vectors*

$$\int_{S(z_i(0))} P_H F_k(0, \cdot) \, dx, \quad \int_{S(z_i(0))} P_H F_l(0, \cdot)) \, dx, \quad \int_{S(z_i(0))} P_H F_m(0, \cdot) \, dx$$

(8.2.4)

are linearly independent.

Theorem 8.5 (Local Controllability in 3D) *Given $u_0 \in V$, under the assumptions of Theorems 6.2–6.3 and Assumption 8.1, the results of Theorem 8.3 hold for the 3D case for three active controls v_k, v_l, and v_m, if*

$$\sum_{i=1}^{n} \int_{S(z_i(0))} P_H F_k(0, \cdot) \, dx, \quad \sum_{i=1}^{n} \int_{S(z_i(0))} P_H F_l(0, \cdot) \, dx, \quad \sum_{i=1}^{n} \int_{S(z_i(0))} P_H F_m(0, \cdot) \, dx$$

(8.2.5)

are linearly independent.

8.2.4   Methodology of Controllability Proofs

The main idea of our proofs below is the same as in Chap. 7, that is, to show that each of the mappings

$$R^2 \ni (v_k, v_l) \ \rightarrow \ z_i(T) \in R^2, \quad R^3 \ni (v_k, v_l, v_m) \ \rightarrow \ z_i(T) \in R^3, \tag{8.2.6}$$

associated with Theorems 8.2 and 8.4, considered on some (open) neighborhood of the origin, is 1-1 and its range contains an open neighborhood of $z_i^*(T)$ for some $T > 0$. To this end, we intend to study the invertibility properties of the respective $[2 \times 2]$- and $[3 \times 3]$-matrices:

$$\left(\frac{dz_i(T)}{dv_k} \Big|_{v_j's=0}, \ \frac{dz_i(T)}{dv_l} \Big|_{v_j's=0} \right)$$

$$\left(\frac{dz_i(T)}{dv_k} \Big|_{v_j's=0}, \ \frac{dz_i(T)}{dv_l} \Big|_{v_j's=0}, \ \frac{dz_i(T)}{dv_m} \Big|_{v_j's=0} \right). \tag{8.2.7}$$

In the above and anywhere below the subscript $v_j's = 0$ indicates that the corresponding expressions are calculated for $v_j = 0$, $j = 1, \ldots, 2n - 3$.

8.3   Derivatives $\frac{\partial u}{\partial v_j}|_{v'_j s=0}$: 2D Case

8.3.1   Auxiliary Notations

Select any index $j \in \{1, \ldots, 2n - 3\}$ and assume that in (8.1.1)–(8.1.5)

$$v_j = h \in R, \quad |h| \leq 1, \quad v_m = 0, \quad m \neq j. \tag{8.3.1}$$

For this selection of parameters, denote

$$z_i(t) = z_i(t; h), \quad z(t) = z(t; h),$$

$$F(t, x) = F(t, x; h), \quad F_j(t, x) = F_j(t, x; h), \quad p(t, x) = p(t, x; h),$$

$$u(t, x) = u(t, x; h) = u_h(t, x), \quad u_*(t, x) = u(t, x; 0), \quad w_h = \frac{u_h - u_*}{h}. \tag{8.3.2}$$

8.3.2   Equation for w_h and its Well-Posedness

Our goal is to study the behavior of w_h as h tends to zero.

To this end we intend to employ the classical Galerkin method along the associated techniques used in the classical theory of parabolic pde's in [38, Chapters VII and III].

From (8.1.1) we derive

$$\begin{cases} w_{ht} = v \, \Delta w_h - (u_* \cdot \nabla)w_h - (w_h \cdot \nabla)u_h \\ \quad + F_j(\cdot, \cdot; h) - \frac{1}{h}\nabla(p(\cdot, \cdot; h) - p(\cdot, \cdot; 0)) & \text{in } (0, T^*) \times \Omega, \\ \text{div } w_h = 0 & \text{in } (0, T^*) \times \Omega, \text{ i.e., } w_h(t, \cdot) \in H, \\ w_h = 0 & \text{in } (0, T^*) \times \partial\Omega, \\ w_h(0, \cdot) = 0 & \text{in } \Omega. \end{cases} \tag{8.3.3}$$

By Theorem 6.1, (8.1.3)–(8.1.5) and due to the continuous embedding (for $\Omega \subset R^2$, see Remark 8.4 below)

$$C([0, T^*]; [L^2(\Omega)]^2) \bigcap L^2(0, T^*; V) \subset [L^4(Q_{T^*})]^2 = [L_{4,4}(Q_{T^*})]^2, \tag{8.3.4}$$

we have

$$u_*, u_h, w_h \in C([0, T^*]; H) \bigcap L^2(0, T^*; V) \bigcap [L^4(Q_{T^*})]^2, \quad F_j(\cdot, \cdot; h) \in [L^\infty(Q_{T^*})]^2. \tag{8.3.5}$$

Remark 8.4 In the above we used estimate (3.4) in [38], page 75, namely:

$$\|\psi\|_{L^4(Q_t)} \leq \beta\left(\max_{\tau\in[0,t]}\|\psi(\tau,\cdot)\|_{L^2(\Omega)} + \|\nabla\psi\|_{[L^2(Q_t)]^2}\right). \tag{8.3.6}$$

Well-Posedness of System (8.3.3) We will show that Theorem 1.1 in [38, pages 573–574] on well-posedness of general parabolic systems (see Sect. 8.3.3 below) implies that (8.3.3) admits a unique solution with the properties regularity described in (8.3.5), and for some constant $C^* > 0$ the following estimate holds:

$$\|w_h\|_{C([0,T^*];[L^2(\Omega)]^2)\cap L^2(0,T^*;V)} \overset{\Delta}{=} \max_{t\in[0,T^*]}\|w_h(t,\cdot)\|_{[L^2(\Omega)]^2}$$

$$+ \|w\|_{L^2(0,T^*;V)} \leq C^*\|PF_j(\cdot,\cdot;h)\|_{[L_{2,1}(Q_{T^*})]^2} \leq C^*\|F_j(\cdot,\cdot;h)\|_{[L_{2,1}(Q_{T^*})]^2}. \tag{8.3.7}$$

Indeed, the proof of this theorem is based on the classical Galerkin methods with the following test functions:

$$\phi \in L^2(0,T^*;[H_0^1(\Omega)]^2)\bigcap\{\phi(\cdot,x) \in [H^1(0,T^*)]^2 \quad \text{a.e. in } \Omega\}.$$

However, in the case of the special mixed problem (8.3.3), including the extra condition

$$\text{div } w_h = 0,$$

we are dealing with $w_h(t,\cdot)$ that lie in V for almost all t, see (8.3.5).

Therefore, w_h can be represented as the following Fourier series, expanded *only* along the eigenfunctions $\{\omega_k\}_{k=1}^\infty, \omega_k \in V, k = 1,\ldots$ of the spectral problem, associated with (8.1.1) (let us remind the reader that these eigenfunctions form a complete orthogonal basis in V and orthonormal in H, see Sect. 7.2.2 in Chap. 7 and [36]):

$$w_h(t,x) = \sum_{k=1}^\infty c_k(t)\omega_k(x); \quad \nu\Delta\omega_k = \lambda_k\omega_k + \nabla p_k, \text{ div } \omega_k = 0 \text{ in } \Omega, \quad k = 1,\ldots. \tag{8.3.8}$$

The equation for ω_k's is understood in the sense of the following identity (see also (8.3.9)):

$$-\nu[w_k,\phi] = \lambda_k\int_\Omega \omega_k\phi dxd\tau \quad \forall\phi \in V.$$

In other words, (8.3.3) is equivalent to the following identity obtained as the difference of identities (8.1.8) in the cases when $u = u_h$ and $u = u_*$ and then divided by h (compare to [38, p. 572]):

$$\int_{\Omega} w_h(x,t)\phi(t)dx - \int_0^t \int_{\Omega} w_h \cdot \phi_t dx dt$$

$$= -\int_0^t \nu \left[w_h(\tau,\cdot), \phi(\tau,\cdot)\right]dt$$

$$+ \int_0^t \int_{\Omega} \left(-(u_* \cdot \nabla)w_h - (w_h \cdot \nabla)u_h + P_H F_j(\tau,x;h)\right)\phi dx d\tau, \quad t \in [0,T^*],$$

(8.3.9)

where ϕ is any function such that $\phi \in \Phi = \{\phi \in L^2(0,T^*;V), \phi_t \in L^2(0,T^*;H)\}$.

The derivation of (8.3.7) in [38] is based on the classical form of identity (8.3.9), namely for any $\phi \in L^2(0,T^*;[H_0^1(\Omega)]^2)$, $\phi_t \in L^2(Q_{T^*})$, which in this case will be applied for $\phi \in \Phi$, and, thus, is the same as just to use the last line in (8.3.9) from the start.

8.3.3 Auxiliary Regularity Results for Parabolic Systems from [38]

Let us recall the following results from [38].

- Theorem 1.1 in [38, page 573] requires that the squared 1-D components of the 2-D vector-function u_* and 1-D components of the 2×2 matrix-function ∇u_h (as the coefficients in (8.3.3)) are elements of the space $L_{q,r}(Q_{T^*})$, where

$$\|\psi\|_{L_{q,r}(Q_{T^*})} = \left(\int_0^{T^*} \left(\int_{\Omega} |\psi|^q dx\right)^{r/q} dt\right)^{1/r}, \quad \frac{1}{r} + \frac{1}{q} = 1, \quad q \in (1,\infty], \ r \in [1,\infty),$$

  while the free term $F_j(\cdot,\cdot;h)$ lies in $L_{q_1,r_1}(Q_{T^*})$ (due to (8.3.9) we ignore the term $-\frac{1}{h}\nabla(p(\cdot,\cdot;h) - p(\cdot,\cdot;0))$ here), where $\frac{1}{r_1} + \frac{1}{q_1} = 1 + \frac{1}{2}$, $q_1 \in (1,2]$, $r_1 \in [1,2)$. We can select $r_1 = 1, q_1 = 2$ and $r = 2 = q$.
- Constant C^* can be selected to be dependent only on

$$\|u_*\|_{C([0,T^*];[L^2(\Omega)]^2) \cap L^2([0,T^*;V)}$$

  or $\|u_*\|_{[L^4(Q_{T^*})]^2}$, and $\|\nabla u_h\|_{[[L^2(Q_{T^*})]^2]^2}$, $|h| \leq 1$, see [38, pages 573–574] and (8.3.1).
- Condition div $w_h = 0$ is not required in Theorem 1.1 in [38, page 573].
- Note that w_h also satisfy the regularity of solutions to (8.1.1).

Based on the above discussion, we can refine (8.3.7) as follows:

$$\|w_h\|_{C([0,T^*];[L^2(\Omega)]^2) \cap L^2([0,T^*;V)} \leq C^* \|F_j(\cdot,\cdot;h)\|_{[L_{2,1}(Q_{T^*})]^2}$$

$$\leq C^* T^* \operatorname{meas}^{1/2}(\Omega) \, \|F_j(\cdot,\cdot;h)\|_{[L^\infty(Q_{T^*})]^2}$$

$$\leq C^* T^* \operatorname{meas}^{1/2}(\Omega) \, C_{\Omega,r}, \quad j = 1, \ldots, 2n-3, \tag{8.3.10}$$

where $C_{\Omega,r}$ depends on Ω, r, see Lemma 5.1.

8.3.4 Auxiliary System of Linear Equations Systems

Introduce the following linear system:

$$
\begin{cases}
\left(\frac{\partial u}{\partial v_j}|_{v'_j s=0}\right)_t = v\,\Delta\left(\frac{\partial u}{\partial v_j}|_{v'_j s=0}\right) - (u_* \cdot \nabla)\left(\frac{\partial u}{\partial v_j}|_{v'_j s=0}\right) \\
\quad -\left(\left(\frac{\partial u}{\partial v_j}|_{v'_j s=0}\right)\cdot\nabla\right)u_* + P_H F_j(\cdot,\cdot;0) - \nabla p_j & \text{in } (0, T^*) \times \Omega, \\[2ex]
\operatorname{div}\left(\frac{\partial u}{\partial v_j}|_{v'_j s=0}\right) = 0 & \text{in } (0, T^*) \times \Omega, \\[2ex]
& \text{i.e., } \left(\frac{\partial u}{\partial v_j}|_{v'_j s=0}\right)(t,\cdot) \in H, \\[2ex]
\left(\frac{\partial u}{\partial v_j}|_{v'_j s=0}\right) = 0 & \text{in } (0, T^*) \times \partial\Omega, \\[2ex]
\left(\frac{\partial u}{\partial v_j}|_{v'_j s=0}\right)(0,\cdot) = 0 & \text{in } \Omega,
\end{cases}
\tag{8.3.11}
$$

where (in the sense of distributions, see also Theorem 6.1),

$$\nabla p_j = -\left(\frac{\partial u}{\partial v_j}|_{v'_j s=0}\right)_t$$

$$+v\,\Delta\left(\frac{\partial u}{\partial v_j}|_{v'_j s=0}\right) - (u_* \cdot \nabla)\left(\frac{\partial u}{\partial v_j}|_{v'_j s=0}\right) - \left(\left(\frac{\partial u}{\partial v_j}|_{v'_j s=0}\right)\cdot\nabla\right)u_* + P_H f_j(\cdot,\cdot;0).$$

Remark 8.5 (On Understanding System (8.3.11)) Due to incompressibility ("divergence-free") condition

$$\operatorname{div}\left(\frac{\partial u}{\partial v_j}|_{v'_j s=0}\right) = 0,$$

similar to (8.3.8), we can represent solution to (8.3.11) as the following series:

$$\left(\frac{\partial u}{\partial v_j}|_{v'_j s=0}\right)(t, x) = \sum_{k=1}^{\infty} d_k(t)\omega_k(x). \tag{8.3.12}$$

Then the argument of the classical theory of parabolic pde's ([38, Chapters VII and III]) can be applied to the "cut-off" form (8.3.12) exactly as it is applied in the case when such condition is absent.

Respectively, exactly as in the aforementioned classical theory, making use of the identity like (8.3.9), we can derive the existence of solution to (8.3.11) in $C([0, T^*]; [L^2(\Omega)]^2) \cap L^2([0, T^*; V)$ satisfying (8.3.14).

8.3.5   Derivatives $\frac{\partial u}{\partial v_j}|_{v'_j s=0}$

Lemma 8.1 *Derivatives*

$$\lim_{h \to 0} w_h \overset{\Delta}{=} \frac{\partial u}{\partial v_j}|_{v'_j s=0}, \quad j = 1, \dots, 2n - 3,$$

where the limit is taken with respect to the $C([0, T^*]; [L^2(\Omega)]^2) \cap L^2([0, T^*; V)$-norm, exist as unique solutions to (8.3.11) and

$$\frac{\partial u}{\partial v_j}|_{v'_j s=0} \in C([0, T^*]; H) \cap L^2(0, T^*; V) \cap [L^4(Q_{T^*})]^2. \tag{8.3.13}$$

As a particular case of (8.3.10), the following estimates hold:

$$\|\frac{\partial u}{\partial v_j}|_{v'_j s=0}\|_{C([0,T^*]; H) \cap L^2(0,T^*; V)}$$

$$\leq C^* T^* \operatorname{meas}^{1/2}(\Omega) \|f_j(\cdot, \cdot; 0)\|_{[L^\infty(Q_{T^*})]^2}$$

$$\leq C^* T^* \operatorname{meas}^{1/2}(\Omega) C_{\Omega, r}, \quad j = 1, \dots, 2n - 3, \tag{8.3.14}$$

where C^* can be selected to be dependent only on $\|u_*\|_{[L^4(Q_{T^*})]^2}$ and $\|\nabla u_*\|_{[L^2(Q_{T^*})]^2}$.

Proof of Lemma 8.1

Step 1 We will adapt here the argument of Theorem 4.5 in [38], page 166 (on continuous dependence of solutions to parabolic pde's on coefficients and free terms) to the case of systems of linear parabolic pde's along Remark 8.5. Namely, denote

$$W_h = w_h - \frac{\partial u}{\partial v_j}|_{v'_j s=0}.$$

Then, we will have the following identity for W_h from (8.3.9):

$$\int_\Omega W_h(x,t)\phi(t)dx - \int_0^t \int_\Omega W_h \cdot \phi_t dxdt$$

$$= -\int_0^t \nu \left[W_h(\tau,\cdot), \phi(\tau,\cdot) \right] dt + \int_0^t \int_\Omega \left(-(u_* \cdot \nabla)W_h - (W_h \cdot \nabla)u_* \right) \phi dxd\tau$$

$$+ \int_0^t \int_\Omega \left(F_h + P_H(F_j(\tau,x;h) - F_j(\tau,x;0)) \right) \phi dxd\tau, \quad t \in [0,T^*], \qquad (8.3.15)$$

where $F_h = (w_h \cdot \nabla)(u_* - u_h)$.

Making use of Sect. 8.3.3 (see also calculations in Step 2 of Sect. 8.5.3 below), we can derive, similar to (8.3.7) and [38, page 167] at:

$$\| W_h \|_{C([0,T^*];[L^2(\Omega)]^2) \cap L^2([0,T^*;V)}$$

$$\leq C^* \| F_j(\cdot,\cdot;h) - F_j(\cdot,\cdot;0) \|_{[L_{2,1}(Q_{T^*})]^2} + C^* \| F_h \|_{[L_{q_2,r_2}(Q_{T^*})]^2}, \qquad (8.3.16)$$

where $q_2 = 2q/(q+1) = 4/3$, $r_2 = 2r/(r+1) = 4/3$ and C^* is from (8.3.14).

In turn, see again (8.5.15) in Sect. 8.5.3 below:

$$\| F_h \|_{[L_{q_2,r_2}(Q_{T^*})]^2} \leq K \| \nabla u_* - \nabla u_h \|_{[[L^2(Q_{T^*})]^2]^2} \| w_h \|_{[L^4(Q_{T^*})]^2}$$

$$\leq K_* \| \nabla u_* - \nabla u_h \|_{[[L^2(Q_{T^*})]^2]^2} T^* \operatorname{meas}^{1/2}(\Omega) C_{\Omega,r}, \qquad (8.3.17)$$

where $K_* > 0$ is some constant and we used (8.5.14) and (8.3.6) to derive the second inequality.

Step 2 Note next that, due to (6.2.11), (6.2.16), (5.5.12), for some constant k_r, depending on r,

$$\| u_* - u_h \|_{C([0,T^*];H)} + \| \nabla u_* - \nabla u_h \|_{[[L^2(Q_{T^*})]^2]^2} \leq D^* \| 0 \cdot F_j(\cdot,\cdot;0) - h F_j(\cdot,\cdot;h) \|_{[L_2(Q_{T^*})]^2}$$

$$\leq D^* \sqrt{T^*} \operatorname{meas}^{1/2}(\Omega) C_{\Omega,r} |h| \to 0 \quad \text{as } h \to 0,$$

$$\| F_j(\cdot,\cdot;h) - F_j(\cdot,\cdot;0) \|_{[L_{2,1}(Q_{T^*})]^2} \leq T^* \operatorname{meas}^{1/2}(\Omega) \| F_j(\cdot,\cdot;h) - F_j(\cdot,\cdot;0) \|_{[L^\infty(Q_{T^*})]^2}$$

$$\leq k_r T^* \operatorname{meas}^{1/2}(\Omega) \| z(\cdot;h) - z(\cdot;0) \|_{[C([0,T^*];R^2)]^n}$$

$$\leq k_r (T^*)^2 \operatorname{meas}(\Omega) \| u_* - u_h \|_{C([0,T^*];H)}$$

$$\leq D^* k_r (T^*)^{2.5} \operatorname{meas}^{3/2}(\Omega) C_{\Omega,r} |h| \to 0 \quad \text{as } h \to 0. \qquad (8.3.18)$$

Step 3 Estimates (8.3.16)–(8.3.18) yield that

$$\|w_h - \frac{\partial u}{\partial v_j}|_{v'_j s=0}\|_{C([0,T^*];H) \cap L^2(0,T^*;V)} \leq \mathscr{C}(r, \Omega, T^*)|h| \to 0 \text{ as } h \to 0,$$

$$(8.3.19)$$

where $\mathscr{C}(r, \Omega, T^*) > 0$ is defined by r, Ω, T^* and

$$\mathscr{C}(r, \Omega, T^*) \to 0 \text{ as } T^* \to 0. \tag{8.3.20}$$

This completes the proof of Lemma 8.1.

Remark 8.6 We would like to note here that the convergence rate in (8.3.19) is linear with respect to $|h|$.

8.4 Derivatives $\frac{\partial z_i}{\partial v_j}|_{v'_j s=0}$ as Solutions to Volterra Equations: 2D Case

8.4.1 Expression for $\frac{z_i(t;h) - z_i(t;0)}{h}$

Let us show that

$$\frac{\partial z_i}{\partial v_j}|_{v'_j s=0} \overset{\triangle}{=} \lim_{h \to 0} \frac{z_i(t; h) - z_i(t; 0)}{h},$$

where the limit is taken in $[C[0, T^*]]^2$-norm exists.

To this end, we will use the integral form of equations for $z_i(t; h)$'s:

$$z_i(t; h) = z_{i,0} + \frac{1}{\text{meas}(S)} \int_0^t \int_{S(z_i(\tau;h))} u(\tau, x; h) \, dx \, d\tau$$

$$= z_{i,0} + \frac{1}{\text{meas}(S)} \int_0^t \int_{S(z_i(\tau;h)) - z_i(\tau;h)} u(\tau, x - z_i(\tau; h); h) \, dx \, d\tau,$$

$$= z_{i,0} + \frac{1}{\text{meas}(S)} \int_0^t \int_{S_i(0,\tau)} u(\tau, x - z_i(\tau; h); h) \, dx \, d\tau,$$

where we denoted, see (5.2.1) in Chap. 5,

$$S_i(0, \tau) = S(z_i(\tau; h)) - z_i(\tau; h), \quad t \in [0, T^*]. \tag{8.4.1}$$

Note that the set $S_i(0, \tau)$ does not depend on h and describes the application of the preassigned engaged geometric control (i.e., rotation in time, see Sect. 5.2) of the set $S(z_i(t))$ *relative to the origin*.

Respectively:

$$\frac{z_i(t; h) - z_i(t; 0)}{h}$$

$$= \frac{1}{\text{meas}(S)} \int_0^t \int_{S_i(0,\tau)} \frac{u(\tau, x - z_i(\tau; h); h) - u(\tau, x - z_i(\tau; 0); 0)}{h} \, dx \, d\tau$$

$$= \frac{1}{\text{meas}(S)} \int_0^t \int_{S_i(0,\tau)} \frac{u(\tau, x - z_i(\tau; h); h) - u(\tau, x - z_i(\tau; 0); h)}{h} \, dx \, d\tau$$

$$+ \frac{1}{\text{meas}(S)} \int_0^t \int_{S_i(0,\tau)} \frac{u(\tau, x - z_i(\tau; 0); h) - u(\tau, x - z_i(\tau; 0); 0)}{h} \, dx \, d\tau.$$

(8.4.2)

In the previous section we studied the integrand in the 2nd term (i.e., w_h and its limit properties as $h \to 0$)), see Lemma 8.1.

8.4.2   Evaluation of the Integrand in the 1st Term on the Right in (8.4.2)

Denote (see the respective expression in (8.4.2)):

$$\rho(s, t, x) \overset{\Delta}{=} (1 - s)(x - z_i(t; 0)) + s(x - z_i(t; h))$$

$$= x - [(1 - s)z_i(t; 0) + s z_i(t; h)] \in \Omega, s \in [0, 1], \; x \in S_i(0, t), \; t \in [0, T^*].$$

We intend to show that (for a.e. t, x):

$$u(t, x - z_i(t; h); h) - u(\tau, x - z_i(t; 0); h)$$

$$= u_x(t, x - z_i(t; 0); 0)(z_i(t; 0)) - z_i(t; h)) \; + \; G(t, x; h)(z_i(t; 0)) - z_i(t; h)),$$

(8.4.3)

where u_x is the Jacobian matrix of the function $u(t, x)$ with respect to x and

$$\| \int_{S_i(0,t)} G(t, x; h) dx \|_{R^2} \leq O(h) \| \nabla u(t, \cdot; h) \|_{[[L^2(\Omega)]^2]^2} \; \text{as } h \to 0, \; t \in [0, T^*].$$

(8.4.4)

Remark 8.7 Here, as usual, when we use a term like $O(p)$ we assume that it may depend on the given parameters in the original problem (such as T^*, Ω, r, u_0, v_i's, selected indices) but the limit property $O(p) \to 0$ as $p \to 0$ holds uniformly over such fixed parameters.

Indeed, for example, if $u = (u_1, u_2)$, $z_i = (z_{i1}, z_{i2})$, then, by the Mean-Value Theorem, applied with respect to the independent variable s, we have

$$u_1(t, \rho(1, t, x); h) - u_1(t, \rho(0, t, x); h) = (u_1(t, x - z_i(t; h); h) - u_1(t, x - z_i(t; 0); h))(1 - 0)$$

$$= u_{1x_1}(t, \rho(s_1, t, x); h)(z_{i1}(t; 0)) - z_{i,1}(t; h)) + u_{1x_2}(t, \rho(s_1, t, x); h)(z_{i2}(t; 0)) - z_{i,2}(t; h)),$$

where $s_1 \in [0, 1]$ and point $\rho(s_1, t, x)$ lies in the line interval connecting points $x - z_i(t; 0)$ and $x - z_i(t; h)$ whose length tends to zero as $h \to 0$. Then,

$$\left| \int_{S_i(0,t)} (u_{1x_1}(t, \rho(s_1, t, x); h) - u_{1x_1}(t, \rho(0, t, x); h)) \, dx \right|$$

$$= \left| \int_{S_i(0,t)} (u_{1x_1}(t, \rho(s_1, t, x); h) - u_{1x_1}(t, x - z_i(t; 0); h)) \, dx \right|$$

$$= \left| \int_{S_i((1-s_1)z_i(t;0)+s_1 z_i(t;h),t)} u_{1x_1}(t, x; h) \, dx - \int_{S_i(z_i(t;0),t)} u_{1x_1}(t, x; h) \, dx \right|$$

$$\leq \left| \int_{S_i((1-s_1)z_i(t;0)+s_1 z_i(t;h),t)\setminus S_i(z_i(t;0),t)} |u_{1x_1}(t, x; h)| \, dx \right|$$

$$+ \int_{S_i(z_i(t;0),t)\setminus S_i((1-s_1)z_i(t;0)+s_1 z_i(t;h),t)} |u_{1x_1}(t, x; h)| \, dx \right|$$

$$\leq \operatorname{meas}(S_i((1 - s_1)z_i(t; 0) + s_1 z_i(t; h), t)\setminus S_i(z_i(t; 0), t))^{1/2} \|u_{1x_1}(t, \cdot; h)\|_{L^2(\Omega)}$$

$$+ \operatorname{meas}(S_i(z_i(t; 0), t)\setminus S_i((1 - s_1)z_i(t; 0) + s_1 z_i(t; h), t)^{1/2} \|u_{1x_1}(t, \cdot; h)\|_{L^2(\Omega)}.$$

This implies (8.4.4), in view of (5.2.4).

8.4.3 Volterra Equations

Combining (8.4.2) and (8.4.3), we obtain the following Volterra equation for $\psi_h = \frac{z_i(\cdot; h) - z_i(\cdot; 0)}{h}$:

$$((I + \mathbb{A} + \mathbb{A}_h)\psi_h)(t) \overset{\Delta}{=} \psi_h(t) + \int_0^t \left(\mathbb{K}_0(t, \tau) + \mathbb{K}_h(t, \tau) \right) \psi_h(\tau) d\tau = g(t; h),$$

$$(8.4.5)$$

$$I + \mathbb{A} + \mathbb{A}_h : [C[0, T^*]]^2 \to [C[0, T^*]]^2,$$

where

$$(\mathbb{A}\psi_h)(t) \overset{\Delta}{=} \int_0^t \mathbb{K}_0(t,\tau)\psi_h(\tau)d\tau,$$

$$(\mathbb{A}_h\psi_h)(t) \overset{\Delta}{=} \int_0^t \mathbb{K}_h(t,\tau)\psi_h(\tau)d\tau,$$

I is the identity operator, and

$$\mathbb{K}_0(t,\tau) = \frac{-1}{\text{meas}(S)} \int_{S_i(0,\tau)} u_x(\tau, x - z_i(\tau;0);0)dx, \quad \mathbb{K}_0 \in L^2((0,T^*) \times (0,T^*)),$$

(8.4.6)

$$\mathbb{K}_h(t,\tau) = \frac{-1}{\text{meas}(S)} \int_{S_i(0,\tau)} G(\tau, x; h)dx, \tag{8.4.7}$$

$$g(t;h) = \frac{1}{\text{meas}(S)} \int_0^t \int_{S_i(0,\tau)} \frac{\partial u}{\partial v_j}|_{v'_j s=0}(\tau, x - z_i(\tau;0))\, dx\, d\tau \; + \; H(t;h),$$

where

$$H(t;h) = \frac{1}{\text{meas}(S)} \int_0^t \int_{S_i(0,\tau)} \frac{u(\tau, x - z_i(\tau;0); h) - u(\tau, x - z_i(\tau;0); 0)}{h}\, dx\, d\tau$$

$$- \frac{1}{\text{meas}(S)} \int_0^t \int_{S_i(0,\tau)} \frac{\partial u}{\partial v_j}|_{v'_j s=0}(\tau, x - z_i(\tau;0))\, dx\, d\tau, \quad \|H(\cdot;h)\|_{C([0,T^*];R^2)} \to 0$$

(8.4.8)

as $h \to 0$, due to Lemma 8.1.

It is well-known that, due to the Open Mapping Theorem, the operator $(I + \mathbb{A} + \mathbb{A}_h)$ in (8.4.5) is bijective and the inverse operator $(I + \mathbb{A} + \mathbb{A}_h)^{-1}$ is bounded.

Selection of T^* *Recall that in (8.3.1) we assumed that $|h| \leq 1$. Without loss of generality, from now on, we can assume that T^* is small enough to ensure (making use of the regularity results in Chaps. 5 and 6) that*

$$\|\mathbb{A}\| < \frac{1}{4}, \quad \|\mathbb{A}_h\| < \frac{1}{4} \quad \forall |h| \leq 1. \tag{8.4.9}$$

Respectively, in this case, there exists a constant M_o such that

$$\|\psi_h\|_{C([0,T^*];R^2)} \leq M_o, \quad |h| \leq 1.$$

In view of (8.4.4), (8.4.8), and (8.3.19), we can pass to the limit in (8.4.5), described as

$$\psi_h = \left(I + \mathbb{A}\right)^{-1} \left(g(t; h) - \mathbb{A}_h \psi_h\right) = \left(\sum_{k=0}^{\infty} \mathbb{A}^k\right)\left(g(t; h) - \mathbb{A}_h \psi_h\right),$$

in the $[C[0, T^*]]^2$-norm as $h \to 0$ to establish the existence of

$$\frac{\partial z_i(t)}{\partial v_j}|_{v'_j s=0} = \lim_{h \to 0} \frac{z_i(t; h) - z_i(t; 0)}{h}$$

as the unique solution to the following limit Volterra equation:

$$\frac{\partial z_i(t)}{\partial v_j}|_{v'_j s=0} = -\int_0^t \left\{\frac{1}{\text{meas}(S)} \int_{S_i(0,\tau)} u_x(\tau, x - z_i(\tau; 0); 0)dx\right\}\left[\frac{\partial z_i(\tau)}{\partial v_j}|_{v'_j s=0}\right] d\tau$$

$$+ \frac{1}{\text{meas}(S)} \int_0^t \int_{S_i(0,\tau)} \frac{\partial u}{\partial v_j}|_{v'_j s=0}(\tau, x - z_i(\tau; 0)) \, dx \, d\tau, \tag{8.4.10}$$

and

$$\|\frac{\partial z_i}{\partial v_j}|_{v'_j s=0}\|_{C([0,T^*]; R^2)}$$

$$\leq \left(\sum_{k=0}^{\infty} \|\mathbb{A}_t^k\|\right) \|\frac{1}{\text{meas}(S)} \int_0^{(\cdot)} \int_{S_i(0,\tau)} \frac{\partial u}{\partial v_j}|_{v'_j s=0}(\tau, x - z_i(\tau; 0)) \, dx \, d\tau\|_{C([0,T^*]; R^2)},$$

$$\leq L_0 t^2 C_{\Omega,r}, \tag{8.4.11}$$

where $L_o > 0$ is some constant, $\mathbb{A}_t$ is calculated as $\mathbb{A}$ for the time interval $(0, t)$ in place of $(0, T^*)$ and where we also used (8.3.14).

Thus, we arrived at the following lemma.

Lemma 8.2 *Assume (8.4.9). Then derivatives* $\frac{\partial z_i}{\partial v_j}|_{v'_j s=0}, i = 1, \ldots, n, j = 1, \ldots, 2n - 3$ *are elements of* $C([0, T^*]; R^2)$ *and satisfy (8.4.10).*

Estimate (8.4.11) immediately yields the following lemma from (8.4.10).

Lemma 8.3 *Assume (8.4.9). Then for any* $j = 1, \ldots, 2n - 3$ *and* $t \in [0, T^*]$:

$$\|\frac{\partial z_i}{\partial v_j}|_{v'_j s=0} - \frac{1}{\text{meas}(S)} \int_0^{(\cdot)} \int_{S(z_i(\tau;0))} \frac{\partial u}{\partial v_j}|_{v'_j s=0}(\tau, x) \, dx \, d\tau\|_{[C[0,t]]^2}$$

$$= t^2 O(t) \text{ as } t \to 0. \tag{8.4.12}$$

8.5   Proofs of Theorems 8.1 and 8.2

8.5.1   Further Modification of (8.4.12)

Let w_j^o stand for solution to (8.3.11) with the right-hand side to be $P_H F_j(0, \cdot; 0)$:

$$
\begin{cases}
w_{jt}^o = \nu \, \Delta w_j^o - (u_* \cdot \nabla) w_j^o \\
\quad - (w_j^o \cdot \nabla) u_* + P_H F_j(0, \cdot; 0) - \nabla p_j & \text{in } (0, T^*) \times \Omega, \\
\text{div } w_j^o = 0 & \text{in } (0, T^*) \times \Omega, \text{ i.e., } w_j^o(t, \cdot) \in H, \\
w_j^o = 0 & \text{in } (0, T^*) \times \partial\Omega, \\
w_j^o(0, \cdot) = 0 & \text{in } \Omega.
\end{cases}
$$

(8.5.1)

Then, due to (8.3.14), we have

$$
\| \frac{\partial u}{\partial v_j} |_{v_j' s = 0} - w_j^o \|_{C([0,t]; H)}
$$

$$
\leq C^* \| F_j(\cdot, \cdot; 0) - F_j(0, \cdot; 0) \|_{[L^2(Q_t)]^2}
$$

$$
\leq \sqrt{C^* \int_0^t \int_{S(z_j(\tau;0)) \cap S(z_j(0;0))} (F_j(\tau, x; 0) - F_j(0, x; 0))^2 dx d\tau}
$$

$$
+ \sqrt{C^* \int_0^t \int_{S(z_j(0;0)) \setminus S(z_j(\tau;0))} (F_j(0, x; 0))^2 dx d\tau}
$$

$$
+ \sqrt{C^* \int_0^t \int_{S(z_j(\tau;0)) \setminus S(z_j(0;0))} (F_j(\tau, x; 0))^2 dx d\tau} = t O(t) \text{ as } t \to 0,
$$

(8.5.2)

where we took into account (8.1.9) in Assumption 8.1.

Combining (8.5.2) with (8.4.12), yields

$$
\| \frac{\partial z_i}{\partial v_j} |_{v_j' s = 0} - \frac{1}{\text{meas}(S)} \int_0^{(\cdot)} \int_{S(z_i(\tau;0))} w_j^o \, dx \, d\tau \|_{[C[0,t]]^2}
$$

$$
= t^2 O(t) \text{ as } t \to 0.
$$

(8.5.3)

8.5.2 Proofs of Theorem 8.1 and of Theorem 8.2 in the Case of Local Controllability Near Equilibrium (i.e., When $u_0 = 0$)

8.5.2.1 Step 1

The equilibrium position for the swimmer in (8.1.1)–(8.1.5) is the pair of solutions $(u = 0 = u_0, z = z(0))$ with the initial datum $u_0 = 0$, $v_i = 0$, $i = 1, \ldots, 2n - 3$ and any set of $z_{i,0}$, $i = 1, \ldots, n$.

In this case, (8.5.1) becomes a system of *linear* non-stationary Stokes equations as follows:

$$
\begin{cases}
w^o_{jt*} = \nu \Delta w^o_{j*} + P_H F_j(0, \cdot; 0) - \nabla p^0_j & \text{in } (0, T^*) \times \Omega, \\
\operatorname{div} w^o_{j*} = 0 & \text{in } (0, T^*) \times \Omega, \text{ i.e., } w^o_j(t, \cdot) \in H, \\
w^o_{j*} = 0 & \text{in } (0, T^*) \times \partial\Omega, \\
w^o_{j*} = 0 & \text{in } \Omega.
\end{cases}
$$

$$(8.5.4)$$

Respectively, its solution is represented by the following Fourier series (see Chap. 7):

$$
w^o_{j*}(t, x) = \sum_{k=1}^{\infty} \int_0^t e^{-\lambda_k(t-\tau)} \left(\int_\Omega F'_j(0, q; 0)\omega_k dq \right) d\tau \, \omega_k(x)
$$

$$
= t \sum_{k=1}^{\infty} \frac{1 - e^{-\lambda_k t}}{t\lambda_k} \left(\int_\Omega (P_H F'_j(0, \cdot; 0))(q)\omega_k dq \right) \omega_k(x),
$$

$$
= t(P_H F'_j(0, \cdot; 0))(x) - t \sum_{k=1}^{\infty} \left(1 - \frac{1 - e^{-\lambda_k t}}{t\lambda_k} \right) \left(\int_\Omega (P_H F'_j(0, \cdot; 0))(q)\omega_k dq \right) \omega_k(x).
$$

8.5.2.2 Step 2

Since

$$
\lim_{s \to 0+} \left(1 - \frac{1 - e^{-s}}{s} \right) = 0,
$$

$$
\lim_{s \to \infty} \left(1 - \frac{1 - e^{-s}}{s} \right) = 1,
$$

and the function $(1 - \frac{1-e^{-s}}{s})$ is strictly monotone increasing on $(0, \infty)$,

$$\|w^o_{j*}(t, \cdot) - t P_H F'_j(0, \cdot; 0)\|_{[L(\Omega)]^2} = t O(t) \text{ as } t \to 0, \tag{8.5.5}$$

where $O(t)$ depends on $F_j(0, x; 0)$ (besides λ_k's, i.e., Ω), namely on the rate of convergence of the series

$$\sum_{k=1}^{\infty} \left(\int_{\Omega} (P_H F'_j(0, \cdot; 0))(q)\omega_k dq \right) \omega_k(x)$$

in H.

Combining (8.5.5) with (8.5.3) implies that the term $t P_H F'_j(0, \cdot; 0)$ will define the direction of vector $\frac{\partial z_i}{\partial v_j}|_{v'_j s=0}$ as $t \to 0$, namely:

$$\|\frac{\partial z_i}{\partial v_j}|_{v'_j s=0} - \frac{t^2}{2\text{meas}(S)} \int_{S(z_i(\tau;0))} (P_H F'_j(0, \cdot; 0))(x)dx\|_{[C[0,t]]^2} = t^2 O(t) \text{ as } t \to 0+ \tag{8.5.6}$$

or

$$z_i(t; h) = z_i(t; 0)$$

$$+ \frac{ht^2}{2\text{meas}(S)} \int_{S(z_i(\tau;0))} (P_H F'_j(0, \cdot; 0))(x)dx + ht^2 \rho_j(t) + h\zeta_j(h, t), \ t \in [0, T^*], \tag{8.5.7}$$

where $\|\rho_j\|_{[C[0,t]]^2} = O(t)$ and $\|\zeta_j(h, \cdot)\|_{[C[0,t]]^2} = O(h)$.

8.5.2.3   Step 3: Proof of Theorem 8.1 when $u_0 = 0$

If we replace (8.3.1) with

$$v_j = ha_j \in R, \ |h| \le 1, \ \sum_{j=1}^{2n-3} a_j^2 = 1, \tag{8.5.8}$$

then we can repeat all the calculations leading to (8.5.7) for the force term

$$F(t, x) = F(t, x; h) = \sum_{j=1}^{2n-3} v_j F_j(t, x) = \sum_{j=1}^{2n-3} v_j F_j(t, x; h),$$

in place of the force term in (8.3.2) and obtain (8.2.1) instead.

This proves Theorem 8.1 when $u_0 = 0$.

8.5.2.4 Step 4

Introduce the following sets:

$$V_{k,l,h} = \{v = (v_k, v_l) \mid (v_k, v_l) = h(a_k, a_l), a_k^2 + a_l^2 = 1\}, \ V_{k,l}^{h_0} = \bigcup_{0 \le h \le h_0} V_{k,l,h}, \ h_0 \in [0, 1].$$

Then, (8.2.1) implies that the mapping

$$\mathscr{A}_{t,k,l} : V_{k,l}^{h_0} \ni v \to z_i(t)$$

admits the following representation:

$$z_i(t; h) = z_i(t; 0)$$

$$+ \frac{t^2}{2 \mathrm{meas}(S)} \left(v_k \int_{S(z_i(\tau;0))} (P_H F_k'(0, \cdot; 0))(x)dx + v_l \int_{S(z_i(\tau;0))} (P_H F_l'(0, \cdot; 0))(x)dx \right)$$

$$+ \|v\|_{R^2} t^2 \rho_{k,l}(t) + \|v\|_{|R^2} \zeta_{k,l}(\|v\|_{R^2}, t), \quad t \in [0, T^*], \tag{8.5.9}$$

where $\|\rho\|_{[C[0,t]]^2} = O(t)$ and $\|\zeta_{k,l}(\|v\|_{R^2}, \cdot)\|_{[C[0,t]]^2} = O(\|v\|_{R^2})$.

If the vectors in (8.2.2) are linear independent and, due to continuity of z_i's with respect to $v_j F_j$'s), we can derive from (8.5.9) that starting from some positive "small" h_0 and for some $T \in (0, T^*]$, the mapping $\mathscr{A}_{T,k,l}$ is continuous, 1-1 and the range set $\mathscr{A}_{T,k,l}(V_{k,l}^{h_0})$ is closed.

This also means that the images of the sets $\mathscr{A}_{T,k,l}(V_{k,l,h}), h \in [0, h_0]$ will be closed curves encircling some neighborhoods of the point $z_i(T; 0)$ and that $z_i(T; 0) = z^*(T)$ is an internal point of the set $\mathscr{A}_{T,k,l}(V_{k,l}^{h_0}) = \bigcup_{h \in [0,h_0]} \mathscr{A}_{T,k,l}(V_{k,l,h})$, which implies the result of Theorem 8.2 in the case of local controllability near equilibrium when $u_0 = 0$.

8.5.3 Proof of Theorems 8.2 and 8.1

8.5.3.1 Step 1

We will prove Theorems 8.2 and 8.1 in the general case by adopting the formula (8.5.5) to the linear system (8.5.1).

To this end, let us split solution to system (8.5.1) into the sum of two functions:

$$w_j^o = w_{j*}^o + u_e,$$

where u_e solves (8.5.1) in the case when $P_H F_j(0, \cdot; 0) = 0$, i.e., for the following free term only (see also Remark 8.5):

$$F^* = (u_* \cdot \nabla)w_j^* - (w_j^* \cdot \nabla)u_*, \tag{8.5.10}$$

$$\begin{cases} u_{et} = \nu \, \Delta u_e + F^* - \nabla p_j^* & \text{in } (0, T^*) \times \Omega, \\ \operatorname{div} u_e = 0 & \text{in } (0, T^*) \times \Omega, \text{ i.e., } u_e(t, \cdot) \in H, \\ u_e = 0 & \text{in } (0, T^*) \times \partial\Omega, \\ u_e = 0 & \text{in } \Omega. \end{cases} \tag{8.5.11}$$

Then, the general case of Theorems 8.2 and 8.1 will follow as in Sect. 8.5.2, if we will show that

$$\|u_e(t, \cdot)\|_{[L^2(\Omega)]^2} = tO(t) \text{ as } t \to 0. \tag{8.5.12}$$

8.5.3.2   Step 2

Let us invoke the results obtained within the proof of Theorems 1.1 in [38, page 573] and Theorem 4.5 in [38, page 166], stating that (see (4.8_1)–(4.8_2) in [38, page 156]):

$$F^* \in [L_{q_1, r_1}(Q_{T^*})]^2, \quad q_1 = \frac{2q}{q+1} = \frac{4}{3}, \quad r_1 = \frac{2r}{r+1} = \frac{4}{3},$$

where $q = r = 2$ are as in Sect. 8.3.3.

This selection of q_1 and r_1 satisfies the assumptions in this remark needed to apply the estimate (8.3.7)/(8.3.14) to u_e. Thus, we obtain that

$$\|u_e\|_{C([0,t];[L^2(\Omega)]^2) \cap L^2([0,t;V)} \leq C_* \|F^*\|_{[L^{4/3}(Q_t)]^2}, \quad t \in (0, T^*]. \tag{8.5.13}$$

Now, making use of Hölder's inequality, namely

$$\left| \int_{Q_t} \psi_1^{4/3} \psi_2^{4/3} dx dt \right| \leq \left(\int_{Q_t} \psi_1^4 dx dt \right)^{1/3} \left(\int_{Q_t} \psi_2^2 dx dt \right)^{2/3}, \tag{8.5.14}$$

and of the estimate (8.3.6), applied to w_j^*, we can show that

$$\|F^*\|_{[L^{4/3}(Q_t)]^2} \leq K \left\{ \|u_*\|_{[L^4(Q_t)]^2} \|\nabla w_j^*\|_{[[L^2(Q_t)]^2]^2} + \|w_j^*\|_{[L^4(Q_t)]^2} \|\nabla u_*\|_{[[L^2(Q_t)]^2]^2} \right\}$$

$$\leq KL \left\{ \|u_*\|_{[L^4(Q_t)]^2} + \|\nabla u_*\|_{[[L^2(Q_t)]^2]^2} \right\} \|w_j^*\|_{C([0,T^*];[L^2(\Omega)]^2) \cap L^2([0,T^*;V)} = tO(t) \tag{8.5.15}$$

as $t \to 0$, for some positive constants K, L. Here, the last equality is due to estimates (8.3.10)/(8.3.14) applied to w_j^*.

Finally, combining (8.5.13) and (8.5.15) yields (8.5.12).

This ends the proof of Theorem 8.2.

8.6 Proofs of Theorems 8.1 and 8.4

In the 3D case, we will need to do the following modifications in the above arguments for the 2D case.

8.6.1 Adjustments in Sects. 8.3 and 8.4

- Recall that due to Theorem 6.2,

$$u_*, u_h, W_h \in C([0, T^*]; H) \bigcap C([0, T^*]; V) \quad F_j(\cdot, \cdot; h) \in [L^\infty(Q_{T^*})]^3. \tag{8.6.1}$$

- Due to embedding (see, e.g., [36], [38]):

$$V \subset [H^1(\Omega)]^3 \subset [L^s(\Omega)]^3, \quad s \in [1, 6),$$

$$u_*, u_h, w_h \in C([0, T^*]; V) \subset C([0, T^*]; [L^s(\Omega)]^3) \subset [L_{s,\rho}(Q_{T^*})]^3, \quad \rho > 0, s \in [1, 6). \tag{8.6.2}$$

- Theorem 1.1 in [38, pages 573] (see Sect. 8.3.3) requires
  1. the squared 1-D components of the 3-D vector-function u_* and
  2. 1-D components of the 3×3 matrix-function ∇u_h (as the coefficients in (8.3.3))

  to lie in $L_{q,r}(Q_{T^*})$, where

$$\frac{1}{r} + \frac{3}{2q} = 1, \quad q \in (1.5, \infty], \quad r \in [1, \infty),$$

- while $F_j(\cdot, \cdot; h)$ (we can ignore $-\frac{1}{h}\nabla(p(\cdot, \cdot; h) - p(\cdot, \cdot; 0))$ due to Remark 8.5) lies in $L_{q_1, r_1}(Q_{T^*})$, where

$$\frac{1}{r_1} + \frac{3}{2q_1} = 1 + \frac{3}{4}, \quad q_1 \in [6/5, 2], \quad r_1 \in [1, 2]. \tag{8.6.3}$$

We can select any suitable r_1, q_1 in the above intervals, since $F_j(\cdot, \cdot; h) \in [L^\infty(Q_T^*))]^3$. In particular, we can select, e.g., $q_1 = 2, r_1 = 1$, to preserve the respective space for the free term in (8.3.7).

For the squared 1-D components of t u_* we can pick $q = 2$ and $r = 4$, due to (8.6.2).

In view of (8.6.2), for the 1-D components of ∇u_h we can also pick $q = 2$ and $r = 4$, implying the embedding

$$\nabla u_*, \nabla u_h \in C([0, T^*]; [[L^2(\Omega)]^3]^3).$$

Thus, the respective results of the previous sections, dealing with the 2D case, hold true in the 3D case as well with C^* in (8.3.14) that can be selected to be dependent only on $\|u_*\|_{C([0,T^*]; V)}$.

8.6.2   Adjustments in Sect. 8.5

The results of Sect. 8.5.2 remain the same in the 3D case up to Step 4, which we can modify as follows.

8.6.2.1   Section 8.5.2.4, Step 4 in the 3D Case

Consider now the following sets for controls:

$$V_{k,l,m,h} = \{v = (v_k, v_l, v_m) = h(a_k, a_l, a_m) \mid a_k^2 + a_l^2 + a_m^2 = 1\},$$

$$V_{k,l}^{h_0} = \bigcup_{0 \le h \le h_0} \mathscr{A}_{t,k,l,m}(V_{k,l,m,h}), h_0 \in [0, 1].$$

Then, if the vectors

$$\int_{S(z_i(0))} (P_H F_k(0, \cdot))(x)\, dx, \int_{S(z_i(0))} (P_H F_l(0, \cdot))(x)\, dx, \int_{S(z_i(0))} (P_H F_m(0, \cdot))(x)\, dx$$

(8.6.4)

are linear independent, similar to the argument in Sect. 8.5.2.4, we can show that for "sufficiently small" $h_0 > 0$ and $T \in (0, T^*]$, the point $z_i(T; 0) = z^*(T)$ is an internal point for the set $\bigcup_{0 \le h \le h_0} \mathscr{A}_{T,k,l,m}(V_{k,l,m,h})$, which implies the result of Theorem 8.2 in the case of local controllability near equilibrium in the 3D case.

8.6.2.2   Section 8.5.3 in the 3D Case

In Sect. 8.5.3 we will need to make the following modifications in Step 2 to obtain (8.5.12):

• Recall the following results, obtained within the proof of Theorems 1.1 in [38, page 573] and Theorem 4.5 in [38, page 166] (see (4.8$_1$)–(4.8$_2$) in [38, page 156]):

$$F^* = (u_* \cdot \nabla)w_j^* - (w_j^* \cdot \nabla)u_* \in [L_{q_1^*, r_1^*}(Q_{T^*})]^3, \quad q_1^* = \frac{2q}{q+1} = \frac{4}{3}, \quad r_1^* = \frac{2r}{r+1} = \frac{8}{5}$$

and condition (8.6.3) holds for these q_1^* and r_1^* (in place of q_1, r_1) with the above-selected $q = 2$ and $r = 4$.

- Due to Lemma 1.1 in [38, pages 59–60] (on the space dual of $L_{q_1, r_1}(Q_t)$) and estimates (1.11)–(1.12) on page 137 in [38], we have

$$\|F^*\|_{[L_{q_1^*, r_1^*}(Q_t)]^3}$$

$$\leq M_o \left\{ \|u_*^2\|_{[L_{q,r}(Q_t)]^3}^{1/2} \|\nabla w_j^*\|_{[[L^2(Q_t)]^3]^3} + \|w_j^*\|_{[L_{\bar{q},\bar{r}}(Q_t)]^3} \|\nabla u_*\|_{[[L_{q,r}(Q_t)]^3]^3} \right\}$$

$$(8.6.5)$$

for some $M_o > 0$, where

$$q = \frac{\bar{q}}{\bar{q} - 2} = 2, \bar{q} = 4 \quad \text{and} \quad r = \frac{\bar{r}}{\bar{r} - 2} = 4, \bar{r} = \frac{8}{3}.$$

Remark 8.8 Let us recall estimate (3.4) in [38, page 75] for 3-D case, namely:

$$\|\psi\|_{L_{q_*, r_*}(Q_t)} \leq \beta \left(\max_{\tau \in [0,t]} \|\psi(\tau, \cdot)\|_{L^2(\Omega)} + \|\nabla \psi\|_{[L^2(Q_t)]^2} \right), \quad (8.6.6)$$

$1/r_* + 3/(2q_*) = 3/4, r_* \in [2, \infty), q_* \in [2, 6]$.

- Due to (8.6.6), for some $\beta_* > 0$:

$$\|w_j^*\|_{[L_{\bar{q},\bar{r}}(Q_t)]^3} = \|w_j^*\|_{[L_{4,8/3}(Q_t)]^3} \leq \beta_* \|w_j^*\|_{C([0,T^*];[L^2(\Omega)]^2) \cap L^2([0,T^*;V)}.$$

Hence, for some $M_o^* > 0$:

$$\|F^*\|_{[L_{q_1^*, r_1^*}(Q_t)]^3}$$

$$\leq M_o^* \left\{ \|u_*^2\|_{[L_{q,r}(Q_t)]^3}^{1/2} \|\nabla w_j^*\|_{[L^2(Q_t)]^3]^3} + \|w_j^*\|_{L^2(0,T^*;V)} \|\nabla u_*\|_{[[L_{q,r}(Q_t)]^3]^3} \right\}.$$

$$(8.6.7)$$

- Estimate (8.6.7) and the 3D version of estimate (8.3.14), applied for w_j^*, yields that, instead of (8.5.13), we have

$$\|u_e\|_{C([0,t];[L^2(\Omega)]^3) \cap L^2([0,t;V)} \leq \hat{C} \left\{ \|F^*\|_{[L_{q_1^*, r_1^*}(Q_t)]^3} \right\} = t O(t) \text{ as } t \to 0.$$

$$(8.6.8)$$

This ends the proof of Theorem 8.4.

Remark 8.9 (Open Questions and Research Perspectives) It seems plausible that the controllability results of Chaps. 7 and 8 can be extended to the case when the Hooke's uncontrolled elastic forces are also present, e.g., along the ideas of [20] (see also [25, Ch. 14]), where we considered a different setup of swimmer's models in a fluid described by the non-stationary Stokes equations.

Part IV
Transformations of Swimmers' Internal Forces Acting in 2D and 3D Incompressible Fluids

In this part of the monograph, we will investigate how *the original internal forces of a swimmer will change, both in magnitude and in direction*, when it is placed into a 2D or 3D incompressible (divergence-free) fluid. This phenomenon can be viewed as the main reason why a swimming locomotion is possible.

Chapter 9
Transformation of Swimmers' Forces Acting in a 2D Incompressible Fluid

In this chapter we will investigate how the original swimmer's internal forces will change, both in magnitude and direction, when it is placed into a 2D incompressible fluid. We are particularly interested in the case when the swimmer's body parts are small rectangles or discs. To this end, we will follow the ideas of [22].

9.1 Main Results

9.1.1 Qualitative Estimates for Forces Acting Upon Small Sets in an Incompressible 2D Fluid

In this section we will extend the Assumption (A5.1) in Chap. 5 on the sets S as follows:

Assumption 9.1 *Everywhere in this chapter we assume, to simplify unnecessary technicalities (such as performing a "shifting" operation, etc.) that S is a set with properties described in Sect. 5.2 in Sect. 5 and, also, that*

$$S \subset \Omega,$$

S is strictly separated from $\partial\Omega$ and its center is the origin. Let $\xi(x)$ be the characteristic function of S.

Our first result is aimed at the investigation of the influence of the action of a particular swimmer's force, supported, say, on the set $S(z_i(t))$, on other parts of swimmer's body $S(z_j(t))$, $j \neq i$.

Theorem 9.1 *Consider an arbitrary vector $b = (b_1, b_2) \in R^2$. Then, for any subset $A \subset \Omega$ of positive measure of diameter $2r$, which lies strictly outside of S and is strictly separated from $\partial\Omega$, we have*

© The Author(s), under exclusive license to Springer Nature Switzerland AG 2021
A. Khapalov, *Bio-Mimetic Swimmers in Incompressible Fluids*, Lecture Notes
in Mathematical Fluid Mechanics, https://doi.org/10.1007/978-3-030-85285-6_9

$$\frac{1}{\text{meas}\{A\}} \int_A (P_H b\xi)(x)dx = O(r), \quad \| O(r) \|_{R^2} \le C\frac{\| b \|_{R^2}}{\hat{d}^4}\,\text{meas}^{1/2}\{S\}$$

$$(9.1.1)$$

as $r \to 0+$, where $C > 0$ is a (generically denoted) constant, P_H is the projection operator from $(L^2(\Omega))^2$ onto H, and $\hat{d}$ is the smallest of two distances from A to S and to $\partial\Omega$.

Remark 9.1 (Discussion of Theorem 9.1) This theorem states that the effect of a force, supported in an incompressible 2D fluid on a set S, on a similarly sized subset of Ω, that is strictly separated from S, is <u>negligible</u> when:
1. the measure of S decreases and
2. the distance between this set and S and to $\partial\Omega$ increases.

Theorem 9.2 *Let Assumption 9.1 hold. Then:*

$$\frac{1}{\text{meas}\{S\}} \int_S (P_H b\xi)(x)dx = b - \frac{1}{\text{meas}\{S\}} \int_S g(x)dx + O(r), \qquad (9.1.2)$$

as $r \to 0+$, where

$$g = (g_1, g_2),$$

$$g_1(x) = \frac{1}{2\pi}b_1 \left(\int_{S\setminus B_h(x)} \frac{-(x_1 - y_1)^2 + (x_2 - y_2)^2}{((x_1 - y_1)^2 + (x_2 - y_2)^2)^2}dy + \pi \right)$$

$$- \frac{1}{\pi}b_2 \int_{S\setminus B_h(x)} \frac{(x_1 - y_1)(x_2 - y_2)}{((x_1 - y_1)^2 + (x_2 - y_2)^2)^2}dy, \qquad (9.1.3)$$

$$g_2(x) = \frac{1}{2\pi}b_2 \left(\int_{S\setminus B_h(x)} \frac{-(x_2 - y_2)^2 + (x_1 - y_1)^2}{((x_1 - y_1)^2 + (x_2 - y_2)^2)^2}dy + \pi \right)$$

$$- \frac{1}{\pi}b_1 \int_{S\setminus B_h(x)} \frac{(x_1 - y_1)(x_2 - y_2)}{((x_1 - y_1)^2 + (x_2 - y_2)^2)^2}dy, \qquad (9.1.4)$$

and $B_h(x) = \{y \mid \| x - y \|_{R^2} < h\}$ is any ball of some radius $h = h(x) > 0$ with center at x *that lies in S.*

9.1.2   Transformations of Forces Acting Upon Small Rectangles in an Incompressible 2D Fluid

Let in addition to Assumption 9.1,

$$S = \{(x = (x_1, x_2) \mid -p < x_1 < p, \ -q < x_2 < q\}, \tag{9.1.5}$$

where p and q are given positive numbers.

Theorem 9.3 *Let $b = (b_1, b_2) \in R^2$ be a given. Let*

$$q, p, q^{1-a}/p \to 0+ \quad for \ some \ a \in (0, 1). \tag{9.1.6}$$

Then

$$\frac{1}{\operatorname{meas}\{S\}} \int_S (P_H b \xi)(x)dx = (b_1, 0) + \Big(O(q^a) + O(q^{1-a}/p) + O(p)\Big) \| b \|_{R^2} \tag{9.1.7}$$

as $q, p, q^{1-a}/p \to 0+$.

9.1.3   Transformations of Forces Acting Upon Small Discs in an Incompressible 2D Fluid

The next result will be established as an immediate consequence of the proof of Theorem 9.3.

Theorem 9.4 *Let $b = (b_1, b_2) \in R^2$ be a given and S be a <u>disk</u> of radius r. Then*

$$\frac{1}{\operatorname{meas}\{S\}} \int_S (P_H b \xi)(x)dx = \frac{1}{2}(b_1, b_2) + O(r) \| b \|_{R^2} \quad as \ r \to 0+. \tag{9.1.8}$$

9.1.4   Interpretation of Theorems 9.3 and 9.4: What Shape of S Is Better for Locomotion?

1. Assume that we have a swimmer, modeled by (7.2.1)–(7.2.4), whose body parts are small narrow rectangles $S(z_i(t))$'s, and, say, $S(z_1(t)$ at time $t = 0$ is a rectangle on Fig. 9.1. Then, based on Remark 9.1 and Theorem 9.1, we can say that

- if a force $b = (b_1, b_2)$ acts upon $z_1(t)$ at time $t = 0$, then, based on the formula (7.2.8) and Remark 7.2 in Chap. 7, describing the momentary movements of points $z_i(t)$'s, this force b will be transformed into a force b^* which approximately retains only one component of the original force which is parallel to a longer side of the rectangle at hand, i.e., into, approximately, the force $(0, b_1)$ on Fig. 9.1.
- Alternatively, we can say that the aforementioned component will dominate in the transformed force.
- Thus, *the point $z_1(t)$ will try to move approximately in the direction $(0, b_1)$.*

2. Assume now that we have a swimmer, modeled by (7.2.1)–(7.2.4), whose body parts are small discs $S(z_i(t))$'s, and, say, $S(z_1(t)$ at time $t = 0$ is a disc on Fig. 9.2. Then, we can say that

- if a force $b = (b_1, b_2)$ acts upon $z_1(t)$ at time $t = 0$, then this force b will be transformed into a force b^* which approximately will retain the direction of the original force but will lose a half of its magnitude, see Fig. 9.2.

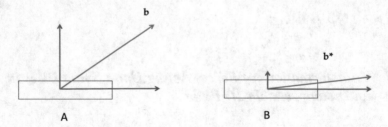

Fig. 9.1 Figure (**a**): a force b acts on a rectangle outside an incompressible medium. Figure (**b**): a force b is transformed into a force b^* if the rectangle is inside an incompressible medium

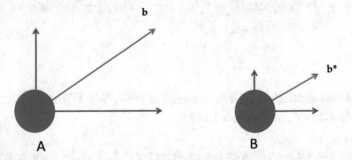

Fig. 9.2 Figure (**a**): a force b acts on a disc outside an incompressible medium. Figure (**b**): a force b is transformed into a force b^* if the disc is inside an incompressible medium

Thus, the sum of the transformed swimmer's forces for a swimmer, constructed of discs, will remain to be zero, and, essentially, *it will not momentarily move the center of swimmer's mass.*

A Qualitative Conclusion on a Desirable Body Shape for Swimmers

- From the above discussion, we can conclude that an asymmetric rectangular shape of swimmer's body parts may allow one to select controls v_l, w_k's to generate, based on the formula (7.2.8) and Remark 7.2 in Chap. 7, a non-zero controlled force that can result in the swimmer's locomotion in the desirable direction. *This line of argument is further explored in Chap. 11.*
- Respectively, the symmetric body parts of a shape of a disc make the controlled locomotion rather questionable.

9.2   Proof of Theorem 9.1

9.2.1   Step 1

Recall first that (Chap. 7 or, e.g., [56]) that $(L^2(\Omega))^2$ is the direct sum of H and $G(\Omega)$, where:

$$H = \{u \in (L^2(\Omega))^2, \ \text{div} \, u = 0, \ \gamma_\nu u \, |_{\partial\Omega} = 0\}, \tag{9.2.1}$$

$$G(\Omega) = \{u \in (L^2(\Omega))^2, \ u = \nabla p, \ p \in H^1(\Omega)\}, \tag{9.2.2}$$

where ν is the unit vector normal to the boundary $\partial\Omega$ (pointing outward) and $\gamma_\nu u \, |_{\partial\Omega}$ is the restriction of $u \cdot \nu$ to $\partial\Omega$.

Hence, we can write:

$$(P_H b\xi)(x) = b\xi(x) - \nabla w(x), \tag{9.2.3}$$

where w solves the following generalized Neumann problem:

$$\Delta w = \ \text{div} \, b\xi(x) \quad \text{in } \Omega, \quad \frac{\partial w}{\partial \nu} \, |_{\partial\Omega} = 0. \tag{9.2.4}$$

Since ξ vanishes in A, (9.2.3) gives

$$P_H b\xi = -\nabla w \quad \text{in } A,$$

and

$$\frac{1}{\text{meas}\,\{A\}} \int_A (P_H b\xi)(x) dx \ = \ -\frac{1}{\text{meas}\,\{A\}} \int_A \nabla w dx. \tag{9.2.5}$$

Thus, to prove Theorem 9.1, it is sufficient to show that

$$\frac{1}{\text{meas}\{A\}}\int_A \nabla w \, dx = O(r) \quad \text{as } r \to 0+. \tag{9.2.6}$$

9.2.2   Step 2: Green's Formula

To derive (9.2.6) we intend to use the generalized version of the classical Green's formula representing solutions of the boundary problems (9.2.4), namely:

$$2\pi w(x) = -\int_{\partial\Omega} w(\eta)\frac{\partial}{\partial v}\left(\ln\frac{1}{\sqrt{(x_1-\eta_1)^2+(x_2-\eta_2)^2}}\right)d\eta$$

$$-\int_{\Omega} \Delta w(y)\left(\ln\frac{1}{\sqrt{(x_1-y_1)^2+(x_2-y_2)^2}}\right)dy$$

$$= -\int_{\partial\Omega} w(\eta)\frac{\partial}{\partial v}\left(\ln\frac{1}{\sqrt{(x_1-\eta_1)^2+(x_2-\eta_2)^2)}}\right)d\eta$$

$$+\int_{S} b^T\nabla\left(\ln\frac{1}{\sqrt{(x_1-y_1)^2+(x_2-y_2)^2}}\right)dy, \quad y=(y_1,y_2). \tag{9.2.7}$$

Remark 9.2 In the above and below, when we write the symbol ∇ within some integral, we mean that the corresponding differentiation is conducted with respect to the integration variables.

To derive (9.2.7), we can, making use of the integration by parts, first establish (9.2.7) for a sequence of solutions to (9.2.4) generated by a sequence of continuously differentiable functions $g_n(x)$ on the right, which converge to $\xi(x)$ in $L^2(\Omega)$, and then pass to the limit as $n \to \infty$.

The 2-nd integral on the right in (9.2.7) is well defined near the "bad point" (x_1, x_2), which can be shown by switching to the polar coordinates near it.

In the above and below, we interpret the improper integral over a given domain E for a function with a discontinuity at x as the limit of the integrals over $E\backslash B_s(x)$ as $s \to 0+$, where $B_s(x)$ is a ball of radius s with center at x.

9.2.3 Step 3: Evaluation of the Integral of the Gradient of the 1-st Terms on the Right in (9.2.7) Over A

Recall first that

$$\| w \|_{L^2(\partial\Omega)} \leq L_0 \| w \|_{H^1(\Omega)},$$

where L_0 depends on the $\partial\Omega$.

Let $\{\alpha_k\}_{k=1}^\infty$ ($\alpha_k \to -\infty$ as $k \to \infty$) and $\{p_k\}_{k=1}^\infty$ be the negative eigenvalues and eigenfunctions, orthonormalized in $L^2(\Omega)$, associated with the spectral problem

$$\Delta p = \alpha p, \quad \frac{\partial p}{\partial \nu} \Big|_{\partial\Omega} = 0.$$

Then, (9.2.4) implies that

$$w(x) = \sum_{k=1}^\infty \frac{1}{\sqrt{-\alpha_k}} \left(\int_\Omega \xi(x) b^T \frac{\nabla p_k}{\sqrt{-\alpha_k}} dx \right) p_k(x) + K,$$

where without loss of generality we can set $K = 0$ (as in (9.2.6) we only deal with ∇w).

Hence, noticing that $\{\frac{\nabla p_k}{\sqrt{-\alpha_k}}\}_{k=1}^\infty$ is an orthonormal sequence in $(L^2(\Omega))^2$, we derive from Bessel's inequality that

$$\| w_1 \|_{H^1(\Omega)} \leq C \| b \|_{R^2} \, \text{meas}^{1/2}\{S\}, \tag{9.2.8}$$

where C stands for a (generic) positive constant.

Furthermore, for $i, j = 1, 2, i \neq j$ and $x \neq y$, we have

$$\frac{\partial}{\partial y_i} \left(\ln \frac{1}{\sqrt{(x_1 - y_1)^2 + (x_2 - y_2)^2}} \right) = \frac{x_i - y_i}{(x_1 - y_1)^2 + (x_2 - y_2)^2}, \tag{9.2.9}$$

$$\frac{\partial^2}{\partial y_i \partial x_i} \left(\ln \frac{1}{\sqrt{(x_1 - y_1)^2 + (x_2 - y_2)^2}} \right) = \frac{-(x_i - y_i)^2 + (x_j - y_j)^2}{((x_1 - y_1)^2 + (x_2 - y_2)^2)^2}, \tag{9.2.10}$$

$$\frac{\partial^2}{\partial y_i \partial x_j} \left(\ln \frac{1}{\sqrt{(x_1 - y_1)^2 + (x_2 - y_2)^2}} \right) = \frac{-2(x_i - y_i)(x_j - y_j)}{((x_1 - y_1)^2 + (x_2 - y_2)^2)^2}. \tag{9.2.11}$$

Let d_0 denote the distance between the set A and $\partial\Omega$:

$$d_0 = \inf_{x \in A, y \in \partial\Omega} \sqrt{(x_1 - y_1)^2 + (x_2 - y_2)^2}. \tag{9.2.12}$$

Combining (9.2.8)–(9.2.11), we can derive that

$$
\| \int_A \nabla \int_{\partial\Omega} w(\eta) \frac{\partial}{\partial\nu} \left(\ln \frac{1}{\sqrt{(x_1-\eta_1)^2+(x_2-\eta_2)^2}} \right) d\eta dx \|_{R^2}
$$

$$
\leq \frac{L_0 C}{d_0^4} \| b \|_{R^2} \ \text{meas}\{A\}\text{meas}^{1/2}\{S\} \ \text{meas}^{1/2}\{\partial\Omega\}, \tag{9.2.13}
$$

where, again, C is a (generic) positive constant.

Remark 9.3 Similarly to the above, we can show that

$$
\| \int_S \nabla \int_{\partial\Omega} w(\eta) \frac{\partial}{\partial\nu} \left(\ln \frac{1}{\sqrt{(x_1-\eta_1)^2+(x_2-\eta_2)^2}} \right) d\eta dx \|_{R^2}
$$

$$
\leq \frac{L_0 C}{d_*^4} \| b \|_{R^2} \ \text{meas}^{3/2}\{S\}\text{meas}^{1/2}\{\partial\Omega\}, \tag{9.2.14}
$$

where d_* denotes the distance between the set S and $\partial\Omega$.

9.2.4 Step 4: Evaluation of the Integral of the Gradient of the 2-nd Term in (9.2.7) Over A

Denote by d_1 the distance between the set S and A:

$$
d_1 = \inf_{x\in S, y\in A} \sqrt{(x_1-y_1)^2+(x_2-y_2)^2}.
$$

Then, similar to (9.2.13), it follows from (9.2.7) that

$$
\| \int_A \nabla \int_S b^T \nabla \left(\ln \frac{1}{\sqrt{(x_1-y_1)^2+(x_2-y_2)^2}} \right) dy dx \|_{R^2}
$$

$$
\leq \frac{C}{d_1^4} \| b \|_{R^2} \ \text{meas}\{S\} \ \text{meas}\{A\}, \tag{9.2.15}
$$

where C is a (generic) positive constant. Combining (9.2.13) and (9.2.15) yields (9.2.6), which provides the result of Theorem 9.1.

9.3 Proof of Theorem 9.2

9.3.1 Step 1

Due to (9.2.3) and similar to (9.2.5), we can derive that

$$\frac{1}{\text{meas}\{S\}} \int_S (P_H b\xi)(x)dx = b - \frac{1}{\text{meas}\{S\}} \int_S \nabla w dx. \tag{9.3.1}$$

In view of Remark 9.3, see (9.2.14), to evaluate the 2-nd term on the right in (9.3.1), it is sufficient to evaluate the integral of the gradient of the 2-nd term in (9.2.7) over S.

9.3.2 Step 2

Pick an arbitrary point $x = (x_1, x_2) \in S$.

Since the set S is open, for any $x \in S$ there exists an $h = h(x) > 0$ such that $B_{h(x)}(x) \subset S$ and $B_{h(x)}(x)$ is strictly separated from ∂S.

From (9.2.7), we obtain

$$\int_S b^T \nabla \left(\ln \frac{1}{\sqrt{(x_1 - y_1)^2 + (x_2 - y_2)^2}} \right) dy$$

$$= \int_{S \setminus B_{h(x)}(x)} b^T \nabla \left(\ln \frac{1}{\sqrt{(x_1 - y_1)^2 + (x_2 - y_2)^2}} \right) dy$$

$$+ \int_{B_{h(x)}(x)} b^T \nabla \left(\ln \frac{1}{\sqrt{(x_1 - y_1)^2 + (x_2 - y_2)^2}} \right) dy.$$

Now note that, in view of (9.2.9):

$$\int_{B_{h(x)}(x)} \left(\ln \frac{1}{\sqrt{(x_1 - y_1)^2 + (x_2 - y_2)^2}} \right)_{y_1} dy$$

$$= \lim_{s \to 0+} \int_{B_{h(x)}(x) \setminus B_s(x)} \frac{x_1 - y_1}{(x_1 - y_1)^2 + (x_2 - y_2)^2} dy$$

$$= - \lim_{s \to 0+} \int_0^{2\pi} \int_s^{h(x)} \cos\theta d\rho d\theta = 0. \tag{9.3.2}$$

The above, and similar calculations, for the integration with respect to y_2 within the circle $B_h(x)$, imply that

$$\int_{B_{h(x)}(x)} b^T \nabla \left(\ln \frac{1}{\sqrt{(x_1 - y_1)^2 + (x_2 - y_2)^2}} \right) dx = 0.$$

Thus,

$$\int_S b^T \nabla \left(\ln \frac{1}{\sqrt{(x_1 - y_1)^2 + (x_2 - y_2)^2}} \right) dy$$

$$= \int_{S \setminus B_{h(x)}(x)} \left(b_1 \frac{x_1 - y_1}{(x_1 - y_1)^2 + (x_2 - y_2)^2} + b_2 \frac{x_2 - y_2}{(x_1 - y_1)^2 + (x_2 - y_2)^2} \right) dy.$$

$$\tag{9.3.3}$$

9.3.3   Step 3

We intend now to calculate the gradient of the 2-nd term in (9.2.7) represented as in (9.3.3).

Fix any $x = (x_1, x_2) \in S$. Due to our selection of h for the given x, for "small" Δx_1:

$$B_{h((x_1+\Delta x_1, x_2))}((x_1 + \Delta x_1, x_2)) \subset S.$$

Furthermore, notice that as in the second equality in (9.3.2):

$$\int_{B_{h((x_1+\Delta x_1, x_2))}((x_1+\Delta x_1, x_2)) \setminus B_{h(x)}((x+\Delta x_1, x_2))} \left(\ln \frac{1}{\sqrt{(x_1 + \Delta x_1 - y_1)^2 + (x_2 - y_2)^2}} \right)_{y_1} dy$$

$$= 0.$$

Taking this into account, we obtain from (9.3.3):

$$\frac{\partial}{\partial x_1} \int\limits_S b^T \nabla \left(\ln \frac{1}{\sqrt{(x_1 - y_1)^2 + (x_2 - y_2)^2}} \right) dy$$

$$= \lim_{\Delta x_1 \to 0} \frac{1}{\Delta x_1}$$

$$\times (\int\limits_{S \setminus B_{h((x_1 + \Delta x_1, x_2))}((x_1 + \Delta x_1, x_2))} b^T \nabla \left(\ln \frac{1}{\sqrt{(x_1 + \Delta x_1 - y_1)^2 + (x_2 - y_2)^2}} \right) dy$$

$$- \int\limits_{S \setminus B_{h(x)}(x)} b^T \nabla \left(\ln \frac{1}{\sqrt{(x_1 - y_1)^2 + (x_2 - y_2)^2}} \right) dy)$$

$$= \lim_{\Delta x_1 \to 0} \frac{1}{\Delta x_1} (\int\limits_{S \setminus B_{h(x)}((x_1 + \Delta x_1, x_2))} b^T \nabla \left(\ln \frac{1}{\sqrt{(x_1 + \Delta x_1 - y_1)^2 + (x_2 - y_2)^2}} \right) dy$$

$$- \int\limits_{S \setminus B_{h(x)}(x)} b^T \nabla \left(\ln \frac{1}{\sqrt{(x_1 - y_1)^2 + (x_2 - y_2)^2}} \right) dy)$$

$$= b_1 \frac{\partial}{\partial x_1} \left(\int\limits_{S \setminus B_h(x)} \frac{x_1 - y_1}{(x_1 - y_1)^2 + (x_2 - y_2)^2} dy \right)$$

$$+ b_2 \frac{\partial}{\partial x_1} \left(\int\limits_{S \setminus B_h(x)} \frac{x_2 - y_2}{(x_1 - y_1)^2 + (x_2 - y_2)^2} dy \right),$$

where $h = h(x)$ in the last line is now treated as independent of x when calculating the derivatives.

9.3.4 Step 4

Let us now calculate the derivative in the 1-st term in the last expression in Step 3. Without loss of generality, we can assume for simplicity that

$$S = \{x \mid -r < x_1 < r, \alpha(x_1) < x_2 < \beta(x_1)\}.$$

To simplify notations, we will further write h instead of $h(x)$.

$$\frac{\partial}{\partial x_1} \int\limits_{S \setminus B_h(x)} \frac{x_1 - y_1}{(x_1 - y_1)^2 + (x_2 - y_2)^2} dy$$

$$= \frac{\partial}{\partial x_1} \{ \int\limits_{-r}^{x_1 - h} \int\limits_{\alpha(y_1)}^{\beta(y_1)} \frac{x_1 - y_1}{(x_1 - y_1)^2 + (x_2 - y_2)^2} dy_2 dy_1$$

$$+ \int\limits_{x_1 + h}^{r} \int\limits_{\alpha(y_1)}^{\beta(y_1)} \frac{x_1 - y_1}{(x_1 - y_1)^2 + (x_2 - y_2)^2} dy_2 dy_1$$

$$+ \int\limits_{x_1 - h}^{x_1 + h} [\int\limits_{\alpha(y_1)}^{x_2 - \sqrt{h^2 - (x_1 - y_1)^2}} \frac{x_1 - y_1}{(x_1 - y_1)^2 + (x_2 - y_2)^2} dy_2$$

$$+ \int\limits_{x_2 + \sqrt{h^2 - (x_1 - y_1)^2}}^{\beta(y_1)} \frac{x_1 - y_1}{(x_1 - y_1)^2 + (x_2 - y_2)^2} dy_2] dy_1 \}$$

$$= \int\limits_{S \setminus B_h(x)} \frac{-(x_1 - y_1)^2 + (x_2 - y_2)^2}{((x_1 - y_1)^2 + (x_2 - y_2)^2)^2} dy + 2 \int\limits_{x_1 - h}^{x_1 + h} \frac{(x_1 - y_1)^2}{h^2 \sqrt{h^2 - (x_1 - y_1)^2}} dy_1$$

$$= \int\limits_{S \setminus B_h(x)} \frac{-(x_1 - y_1)^2 + (x_2 - y_2)^2}{((x_1 - y_1)^2 + (x_2 - y_2)^2)^2} dy + \pi. \qquad (9.3.4)$$

Similar calculations also yield

$$\frac{\partial}{\partial x_2} \int\limits_{S \setminus B_h(x)} \frac{x_2 - y_2}{(x_1 - y_1)^2 + (x_2 - y_2)^2} dy$$

$$= \int\limits_{S \setminus B_h(x)} \frac{-(x_2 - y_2)^2 + (x_1 - y_1)^2}{((x_1 - y_1)^2 + (x_2 - y_2)^2)^2} dy + \pi, \qquad (9.3.5)$$

$$\frac{\partial}{\partial x_j} \int\limits_{S \setminus B_h(x)} \frac{x_i - y_i}{(x_1 - y_1)^2 + (x_2 - y_2)^2} dy$$

$$= - \int\limits_{S \setminus B_h(x)} 2 \frac{(x_i - y_i)(x_j - y_j)}{((x_1 - y_1)^2 + (x_2 - y_2)^2)^2} dy, \quad i \neq j, i, j = 1, 2. \tag{9.3.6}$$

Formulas (9.3.4)–(9.3.6) provide the formula for g in (9.1.3) and (9.1.4) (see Step 1 of the proof). This completes the proof of Theorem 9.2.

9.4   Proofs of Theorems 9.3 and 9.4

9.4.1   Proof of Theorem 9.3

Without loss of generality, we can assume that $q < p$ in (9.1.5).

In view of (9.1.2), to establish (9.1.7), it is sufficient to evaluate the integrals in (9.1.3) and (9.1.4) for the case when S is as in (9.1.5).

Select $B_h(x)$ in Theorem 9.2 as follows:

$$B_{h(x)}(x) = \{y \mid \| x - y \|_{R^2} < \frac{1}{2} \min\{\| x \|_{R^2}, q - \| x \|_{R^2}\} = h(x)\} \subset S.$$

9.4.2   Step 1

We begin with the 1-st term on the right in (9.1.3):

$$\int\limits_{S \setminus B_h(x)} \frac{-(x_1 - y_1)^2 + (x_2 - y_2)^2}{((x_1 - y_1)^2 + (x_2 - y_2)^2)^2} dy = \int\limits_{-p}^{x_1 - h} \int\limits_{-q}^{q} \frac{-(x_1 - y_1)^2 + (x_2 - y_2)^2}{((x_1 - y_1)^2 + (x_2 - y_2)^2)^2} dy$$

$$+ \int\limits_{x_1 + h}^{p} \int\limits_{-q}^{q} \frac{-(x_1 - y_1)^2 + (x_2 - y_2)^2}{((x_1 - y_1)^2 + (x_2 - y_2)^2)^2} dy$$

$$+ \int\limits_{x_1 - h}^{x_1 + h} \int\limits_{-q}^{x_2 - \sqrt{h^2 - (x_1 - y_1)^2}} \frac{-(x_1 - y_1)^2 + (x_2 - y_2)^2}{((x_1 - y_1)^2 + (x_2 - y_2)^2)^2} dy$$

$$+ \int\limits_{x_1 - h}^{x_1 + h} \int\limits_{x_2 + \sqrt{h^2 - (x_1 - y_1)^2}}^{q} \frac{-(x_1 - y_1)^2 + (x_2 - y_2)^2}{((x_1 - y_1)^2 + (x_2 - y_2)^2)^2} dy. \tag{9.4.1}$$

9.4.3　Step 2

Making use of (9.2.9)–(9.2.11), we can further derive from (9.4.1) (with $\tan^{-1} = $ arctan):

$$
\int_{-p}^{x_1-h} \int_{-q}^{q} \frac{-(x_1 - y_1)^2 + (x_2 - y_2)^2}{((x_1 - y_1)^2 + (x_2 - y_2)^2)^2} dy + \int_{x_1+h}^{p} \int_{-q}^{q} \frac{-(x_1 - y_1)^2 + (x_2 - y_2)^2}{((x_1 - y_1)^2 + (x_2 - y_2)^2)^2} dy
$$

$$
= \int_{-q}^{q} \frac{y_1 - x_1}{(x_1 - y_1)^2 + (x_2 - y_2)^2} \Big|_{y_1=-p}^{y_1=x_1-h} dy_2 + \int_{-q}^{q} \frac{y_1 - x_1}{(x_1 - y_1)^2 + (x_2 - y_2)^2} \Big|_{y_1=x_1+h}^{y_1=p} dy_2
$$

$$
= \int_{-q}^{q} \frac{-h}{h^2 + (y_2 - x_2)^2} dy_2 + \int_{-q}^{q} \frac{p + x_1}{(p + x_1)^2 + (y_2 - x_2)^2} dy_2
$$

$$
+ \int_{-q}^{q} \frac{p - x_1}{(p - x_1)^2 + (y_2 - x_2)^2} dy_2 - \int_{-q}^{q} \frac{h}{h^2 + (y_2 - x_2)^2} dy_2
$$

$$
= -\tan^{-1} \frac{y_2 - x_2}{h} \Big|_{y_2=-q}^{y_2=q} + \tan^{-1} \frac{y_2 - x_2}{p + x_1} \Big|_{y_2=-q}^{y_2=q}
$$

$$
+ \tan^{-1} \frac{y_2 - x_2}{p - x_1} \Big|_{y_2=-q}^{y_2=q} - \tan^{-1} \frac{y_2 - x_2}{h} \Big|_{y_2=-q}^{y_2=q}
$$

$$
= -\tan^{-1} \frac{q - x_2}{h} + \tan^{-1} \frac{-q - x_2}{h} + \tan^{-1} \frac{q - x_2}{p + x_1} - \tan^{-1} \frac{-q - x_2}{p + x_1}
$$

$$
+ \tan^{-1} \frac{q - x_2}{p - x_1} - \tan^{-1} \frac{-q - x_2}{p - x_1} - \tan^{-1} \frac{q - x_2}{h} + \tan^{-1} \frac{-q - x_2}{h}. \qquad (9.4.2)
$$

9.4.4　Step 3

Similarly to the derivation of (9.4.2), for the remaining two terms in (9.4.1) we have the following:

$$
\int_{x_1-h}^{x_1+h} \int_{-q}^{x_2-\sqrt{h^2-(x_1-y_1)^2}} \frac{-(x_1 - y_1)^2 + (x_2 - y_2)^2}{((x_1 - y_1)^2 + (x_2 - y_2)^2)^2} dy
$$

$$+ \int_{x_1-h}^{x_1+h} \int_{x_2+\sqrt{h^2-(x_1-y_1)^2}}^{q} \frac{-(x_1-y_1)^2 + (x_2-y_2)^2}{((x_1-y_1)^2 + (x_2-y_2)^2)^2} dy$$

$$= - \int_{x_1-h}^{x_1+h} \frac{y_2-x_2}{(x_1-y_1)^2 + (x_2-y_2)^2} \Big|_{-q}^{x_2-\sqrt{h^2-(x_1-y_1)^2}} dy_1$$

$$- \int_{x_1-h}^{x_1+h} \frac{y_2-x_2}{(x_1-y_1)^2 + (x_2-y_2)^2} \Big|_{x_2+\sqrt{h^2-(x_1-y_1)^2}}^{q} dy_1$$

$$= \int_{x_1-h}^{x_1+h} \frac{\sqrt{h^2-(x_1-y_1)^2}}{h^2} dy_1 + \int_{x_1-h}^{x_1+h} \frac{-q-x_2}{(x_1-y_1)^2 + (x_2+q)^2} dy_1$$

$$- \int_{x_1-h}^{x_1+h} \frac{q-x_2}{(x_1-y_1)^2 + (q-x_2)^2} dy_1 + \int_{x_1-h}^{x_1+h} \frac{\sqrt{h^2-(x_1-y_1)^2}}{h^2} dy_1$$

$$= \pi - \tan^{-1} \frac{h}{q+x_2} + \tan^{-1} \frac{-h}{q+x_2} - \tan^{-1} \frac{h}{q-x_2} + \tan^{-1} \frac{-h}{q-x_2},$$
$$\tag{9.4.3}$$

where we took into account that

$$\int_{x_1-h}^{x_1+h} \frac{\sqrt{h^2-(x_1-y_1)^2}}{h^2} dy_1 = \frac{\pi}{2}.$$

9.4.5 Step 4

Due to the equality

$$\tan^{-1} s + \tan^{-1} \frac{1}{s} = \frac{\pi}{2} \quad \forall s > 0 \ (\tan^{-1} s \in (0, \pi/2)),$$

(9.4.2) and (9.4.3) yield for the expression in (9.4.1):

$$\int_{S \setminus B_h(x)} \frac{-(x_1-y_1)^2 + (x_2-y_2)^2}{((x_1-y_1)^2 + (x_2-y_2)^2)^2} dy =$$

$$= -\pi + \tan^{-1}\frac{q - x_2}{p + x_1} + \tan^{-1}\frac{q + x_2}{p + x_1} + \tan^{-1}\frac{q - x_2}{p - x_1} + \tan^{-1}\frac{q + x_2}{p - x_1}.$$
$$(9.4.4)$$

To proceed further with our proof, we will evaluate (see (9.1.2) and the 1-st term on the right in (9.1.3)) as follows:

$$\frac{1}{\text{meas}\{S\}}\int_S\left(-\pi + \tan^{-1}\frac{q - x_2}{p + x_1} + \tan^{-1}\frac{q + x_2}{p + x_1} + \tan^{-1}\frac{q - x_2}{p - x_1} + \tan^{-1}\frac{q + x_2}{p - x_1}\right)dx.$$
$$(9.4.5)$$

9.4.6   Step 5

Consider now the 2-nd term in (9.4.5) (under the sign of integral).
   Note first that

$$\tan^{-1}(s) \in (-\frac{\pi}{2}, \frac{\pi}{2}) \quad \forall s. \qquad (9.4.6)$$

Denote

$$A(p, q) = \{(x = (x_1, x_2) \mid -q < x_2 < q, \ -p + q^{1-a} < x_1 < p - q^{1-a}\}.$$

Since in Theorem 9.3

$$a \in (0, 1), \ q, p, q^{1-a}/p \ \rightarrow \ 0+,$$

we can assume that $A(p, q) \subset S$ with

$$\text{meas } A(p, q) = 4q(p - q^{1-a}) = 4pq(1 - q^{1-a}/p). \qquad (9.4.7)$$

Furthermore, for $x \in A(p, q)$

$$0 < \frac{q - x_2}{p + x_1} < q^a, \qquad (9.4.8)$$

hence,

$$\tan^{-1}\frac{q - x_2}{p + x_1} \leq \tan^{-1}q^a, \quad x \in A(p, q). \qquad (9.4.9)$$

Making use of (9.4.6)–(9.4.9), we can derive that

$$\left| \frac{1}{\text{meas}\,\{S\}} \int_S \tan^{-1} \frac{q - x_2}{p + x_1} dx \right| = \frac{1}{4pq} \int_{-p}^{p} \int_{-q}^{q} \tan^{-1} \frac{q - x_2}{p + x_1} dx_2 dx_1$$

$$= \frac{1}{4pq} \int_{A(p,q)} \tan^{-1} \frac{q - x_2}{p + x_1} dx + \int_{S \setminus A(p,q)} \tan^{-1} \frac{q - x_2}{p + x_1} dx$$

$$\leq (1 - q^{1-a}/p) \tan^{-1} q^a + \frac{4q^{2-a}}{4pq} \frac{\pi}{2} = O(\tan^{-1} q^a) + O(q^{1-a}/p) \text{ as } p, q^{1-a}/p \to 0+.$$

(9.4.10)

9.4.7 Step 6

In a similar way we can evaluate the remaining terms in (9.4.5), which will result in:

$$\frac{1}{\text{meas}\,\{S\}} \int_{S_0} \left(\int_{S \setminus B_h(x)} \frac{-(x_1 - y_1)^2 + (x_2 - y_2)^2}{((x_1 - y_1)^2 + (x_2 - y_2)^2)^2} dy \right) dx = \frac{1}{\text{meas}\,\{S\}}$$

$$\times \int_S \left(-\pi + \tan^{-1} \frac{q - x_2}{p + x_1} + \tan^{-1} \frac{q + x_2}{p + x_1} + \tan^{-1} \frac{q - x_2}{p - x_1} + \tan^{-1} \frac{q + x_2}{p - x_1} \right) dx$$

$$= -\pi + O(q^a) + O(q^{1-a}/p) \text{ as } q, p, q^{1-a}/p \to 0+. \quad (9.4.11)$$

9.4.8 Step 7

Due to the antisymmetry of the linear functions and our "symmetric" choice of $h(x)$ in the beginning of this proof (i.e., $h((x_1, x_2)) = h((-x_1, x_2)) = h((x_1, -x_2)) = h((-x_1, -x_2))$):

$$\int_S \int_{S \setminus B_h(x)} \frac{(x_i - y_i)(x_j - y_j)}{((x_1 - y_1)^2 + (x_2 - y_2)^2)^2} dy dx = 0, \quad i, j = 1, 2, \ i \neq j.$$

(9.4.12)

9.4.9   Step 8

We now need to evaluate the remaining terms in (9.1.3) and (9.1.4). To this end, similar to (9.4.4) and (9.4.5), for the 1-st term on the right in (9.1.4) we have from (9.4.11):

$$\frac{1}{\text{meas}\{S\}} \int\limits_{S} \left(\int\limits_{S_0 \setminus B_h(x)} \frac{-(x_2 - y_2)^2 + (x_1 - y_1)^2}{((x_1 - y_1)^2 + (x_2 - y_2)^2)^2} dy \right) dx$$

$$= \pi + O(q^a) + O(q^{1-a}/p) \text{ as } q, p, q^{1-a}/p \to 0+. \tag{9.4.13}$$

9.4.10   Step 9

Equations (9.4.11), (9.4.12) and (9.4.13) yield that in (9.1.3) and (9.1.4) for our S we have

$$\frac{1}{\text{mes}\{S\}} \int\limits_{S} g_1 dx = O(q^a) + O(q^{1-a}/p), \tag{9.4.14}$$

$$\frac{1}{\text{mes}\{S\}} \int\limits_{S} g_2 dx = b_2 + O(q^a) + O(q^{1-a}/p), \tag{9.4.15}$$

(9.4.14), (9.4.15) and (9.1.2) yield (9.1.7), which completes the proof of Theorem 9.3.

9.4.11   Proof of Theorem 9.4: Forces Acting Upon Small Discs in a Fluid

It follows immediately from Theorem 9.2, because, due to the symmetry of a disc, all the integrals in (9.1.3) are equal to zero.

Remark 9.4 (Open Questions and Research Perspectives) In this chapter, the general formula for

$$\frac{1}{\text{meas}\{S\}} \int\limits_{S} (P_H b\xi)(x) dx$$

in Theorem 9.2 was used to obtain specific results for the case when S is a rectangle. Based on the aforementioned general theorem, one can similarly investigate other geometric shapes of S (e.g., ellipses) to be used in the design of various bio-mimetic swimmers, or even the sets S that represent the geometric shapes of various living organisms, see also Remarks 10.7, 11.3.

Chapter 10
Transformation of Swimmers' Forces Acting in a 3D Incompressible Fluid

In this chapter we will extend the results of Chap. 9 to the swimmers in an incompressible (divergence-free) 3D fluid. We are particularly interested in the force transformation phenomenon in the case of swimmers whose bodies consist of parallelepipeds and balls. To this end, we will follow the schemes of proofs in Chap. 9, extending them to the 3D case along the ideas of [32].

10.1 Main Results

Assumption 10.1 *Everywhere in this chapter we assume, to simplify unnecessary technicalities (such as performing a "shifting" operation, etc.), that S is a set with properties described in Sect. 5.2 and, also, that*

$$S \subset \Omega,$$

S is strictly separated from $\partial\Omega$ and its center is the origin, and ξ denotes the characteristic function of S.

10.1.1 Qualitative Estimates for Forces Acting Upon Small Sets in an Incompressible 3D Fluid

Our first result is aimed at the investigation of the influence of the action of a particular swimmer's force, supported, say, on the set $S(z_i(t))$, on other parts of swimmer's body $S(z_j(t))$, $j \neq i$ (compare to Theorem 9.1).

Theorem 10.1 *Let $b = (b_1, b_2, b_3) \in R^3$ be given and a set S be as in Assumption 10.1. Then for any subset A of Ω of positive measure which lies strictly*

© The Author(s), under exclusive license to Springer Nature Switzerland AG 2021   139
A. Khapalov, *Bio-Mimetic Swimmers in Incompressible Fluids*, Lecture Notes
in Mathematical Fluid Mechanics, https://doi.org/10.1007/978-3-030-85285-6_10

outside of S and is strictly separated from $\partial\Omega$, we have

$$\left\| \frac{1}{\text{meas}\{A\}} \int_A (P_H b\xi)(x)dx \right\|_{R^3}$$

$$\leq \frac{C \, \|b\|_{R^3}}{d_\star^3} \left(\text{meas}^{1/2}\{\partial\Omega\} + \text{meas}^{1/2}\{S\} \right) \text{meas}\{A\} \, \text{meas}^{1/2}\{S\}, \quad (10.1.1)$$

where $C > 0$ is a (generic) constant, $\xi(x)$ is the characteristic function of S, and $d_\star$ is the smallest out of the distances from A to S and from A to $\partial\Omega$.

Remark 10.1 (Interpretation of Theorem 10.1) This theorem states that the effect of a force, supported in an incompressible 3D fluid on a set S, on a similarly sized subset of Ω, which is strictly separated from S, is *negligible* when
1. the measure of S decreases and
2. the distance between this set and S and to $\partial\Omega$ increases.

10.1.2   A General Formula for $\frac{1}{\text{meas}\{S\}} \int_S (P_H b\xi)(x)dx$

Theorem 10.2 (Compare to Theorem 9.2) *Let $b = (b_1, b_2, b_3) \in R^3$ be given. Then,*

$$\frac{1}{\text{meas}\{S\}} \int_S (P_H b\xi)(x)dx = b - \frac{1}{\text{meas}\{S\}} \int_S g(x)dx + O(p) \, \|b\|_{R^3} \quad \text{as} \quad p \to 0,$$
$$(10.1.2)$$

where $g = (g_1, g_2, g_3)$ and

$$g_1(x) = \frac{b_1}{3} + \frac{b_1}{4\pi} \int_{S\setminus B_h(x)} \frac{(y_2 - x_2)^2 + (y_3 - x_3)^2 - 2(y_1 - x_1)^2}{\sqrt{(y_1 - x_1)^2 + (y_2 - x_2)^2 + (y_3 - x_3)^2}^5} dy$$

$$+ \frac{b_2}{4\pi} \int_{S\setminus B_h(x)} \frac{-3(y_1 - x_1)(y_2 - x_2)}{\sqrt{(y_1 - x_1)^2 + (y_2 - x_2)^2 + (y_3 - x_3)^2}^5} dy$$

$$+ \frac{b_3}{4\pi} \int_{S\setminus B_h(x)} \frac{-3(y_1 - x_1)(y_3 - x_3)}{\sqrt{(y_1 - x_1)^2 + (y_2 - x_2)^2 + (y_3 - x_3)^2}^5} dy, \quad (10.1.3)$$

$$g_2(x) = \frac{b_2}{3} + \frac{b_2}{4\pi} \int_{S\setminus B_h(x)} \frac{(y_3 - x_3)^2 + (y_1 - x_1)^2 - 2(y_2 - x_2)^2}{\sqrt{(y_1 - x_1)^2 + (y_2 - x_2)^2 + (y_3 - x_3)^2}^5} dy$$

$$+ \frac{b_3}{4\pi} \int_{S \setminus B_h(x)} \frac{-3(y_2 - x_2)(y_3 - x_3)}{\sqrt{(y_1 - x_1)^2 + (y_2 - x_2)^2 + (y_3 - x_3)^2}^5} dy$$

$$+ \frac{b_1}{4\pi} \int_{S \setminus B_h(x)} \frac{-3(y_2 - x_2)(y_1 - x_1)}{\sqrt{(y_1 - x_1)^2 + (y_2 - x_2)^2 + (y_3 - x_3)^2}^5} dy, \tag{10.1.4}$$

$$g_3(x) = \frac{b_3}{3} + \frac{b_3}{4\pi} \int_{S \setminus B_h(x)} \frac{(y_1 - x_1)^2 + (y_2 - x_2)^2 - 2(y_3 - x_3)^2}{\sqrt{(y_1 - x_1)^2 + (y_2 - x_2)^2 + (y_3 - x_3)^2}^5} dy$$

$$+ \frac{b_1}{4\pi} \int_{S \setminus B_h(x)} \frac{-3(y_3 - x_3)(y_1 - x_1)}{\sqrt{(y_1 - x_1)^2 + (y_2 - x_2)^2 + (y_3 - x_3)^2}^5} dy$$

$$+ \frac{b_2}{4\pi} \int_{S \setminus B_h(x)} \frac{-3(y_3 - x_3)(y_2 - x_2)}{\sqrt{(y_1 - x_1)^2 + (y_2 - x_2)^2 + (y_3 - x_3)^2}^5} dy. \tag{10.1.5}$$

In the above $B_h(x) = \{y : \|y - x\|_{R^3} < h\}$ *is an arbitrary ball of radius* $h = h(x) > 0$ *with center at* x *that lies in* S.

10.1.3 The Case of Parallelepipeds

The next result deals with parallelepipeds, namely, when

$$S = \{x = (x_1, x_2, x_3) \mid -p < x_1 < p, -q < x_2 < q, -\ell < x_3 < \ell\} \subset \Omega, \quad 0 < \ell \leq q \leq p. \tag{10.1.6}$$

We will consider the following two cases:

1. $\ell << q \leq p$;
2. $\ell \leq q << p$.

Theorem 10.3 (The Case $\ell << q \leq p$) *Let* $b = (b_1, b_2, b_3) \in R^3$ *be given and* S *be as in (10.1.6), where* $0 < \ell < q \leq p$. *Assume that, for some* $\varepsilon \in (0, 1)$,

$$\frac{\ell^{1-\varepsilon}}{q} \to 0 \text{ as } p \to 0.$$

Then,

$$\frac{1}{\text{meas}\{S\}} \int_S (P_H b\xi)(x) dx = (b_1, b_2, 0) + \left[O(\ell^\varepsilon) + O(\frac{\ell^{1-\varepsilon}}{q}) + O(p) \right] \|b\|_{R^3} \tag{10.1.7}$$

as $p \to 0$.

Theorem 10.3 is illustrated in Fig. 10.1.

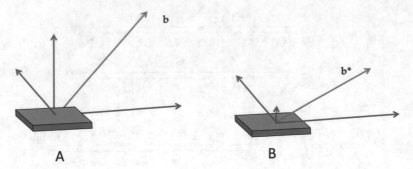

Fig. 10.1 Illustration to Theorem 10.3: in figure (**A**) a force b acts on a parallelepiped outside of an incompressible medium, while this force is transformed into a force b^* in figure (**B**) when the parallelepiped is inside of an incompressible medium

Now we will consider parallelepipeds that have two equal dimensions substantially smaller than the third one.

Theorem 10.4 (The Case $\ell \leq q << p$) *Let $b = (b_1, b_2, b_3) \in R^3$ be given and S be as in (10.1.6), where $0 < \ell < q \leq p$. Assume that, for some $\tau \in (0, 1)$,*

$$\frac{q^{1-\tau}}{p} \to 0 \ as \ p \to 0.$$

Then,

$$\frac{1}{\text{meas}\{S\}} \int_S (P_H b\xi)(x)dx = (b_1, \frac{b_2}{2}, \frac{b_3}{2}) + \left[O(q^\tau) + O(\frac{q^{1-\tau}}{p}) + O(p) \right] \|b\|_{R^3}$$

(10.1.8)

as $p \to 0$.

Theorem 10.4 is illustrated in Fig. 10.2.

Remark 10.2 (Interpretation of (10.1.8)) The result of Theorem 10.4 is somewhat of a surprise. Namely, one could expect that, similar to (9.1.7) in the 2D case, the first term on the right in (10.1.8) should look like $(b_1, 0, 0)$, that is, the parallelepiped would move *predominantly perpendicular to its smallest cross-section—vector* $(b_1, 0, 0)$. To this end, a possible explanation can be the fact that we are dealing with odd dimensional 3D space in this theorem and other intersections of the parallelepiped by hyperplanes contain its largest dimension p.

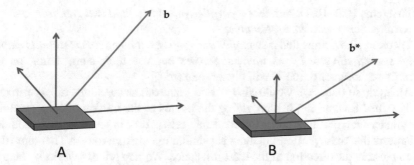

Fig. 10.2 Illustration to Theorem 10.4: in figure (**A**) a force b acts on a parallelepiped outside of an incompressible medium, while this force is transformed into a force b^* in figure (**B**) when the parallelepiped is inside of an incompressible medium

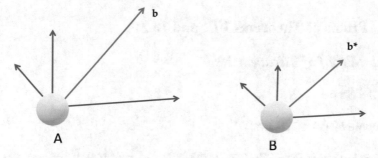

Fig. 10.3 Illustration to Theorem 10.5: in figure (**A**) a force b acts on a ball outside of an incompressible medium, while this force is transformed into a force b^* in figure (**B**) when the ball is inside of an incompressible medium

10.1.4 Spheres in 3D

Theorem 10.5 (The Case of Spheres) *Let* $b = (b_1, b_2, b_3) \in R^3$ *be a given,* $S = B_r(0)$ *(Fig. 10.3). Then,*

$$\frac{1}{\text{meas}\{S\}} \int_S (P_H b\xi)(x)dx = \frac{2}{3}(b_1, b_2, b_3) + O(r)\|b\|_{R^3} \quad \text{as } r \to 0^+. \quad (10.1.9)$$

10.1.5 Instrumental Observations in Relation to Controlled Steering

In view of Theorem 10.1 and Remark 10.1, for swimmers with small body parts, we have the following observations:

- Theorems 10.3–10.4 claim that *a parallelepiped in a fluid will generally move in the direction of the least resistance.*
- Theorem 10.5 claims that *a swimmer that consists of identical small balls cannot be steered directly by its internal forcers* because their sum, acting on the swimmer's center of mass, will always be zero.
- Along these lines, we would also like to point out at the series of experiments described by Leal in [39], providing the photos of the motion of long (rounded cylinder-like) particles under the constant gravity force in several types of fluid. It appears that these particles (somewhat similar to long thin rods in Theorem 10.4) do move in the direction of the least resistance. We also refer to works by Happel and Brenner [16], and by Galdi [11] (see also the references therein), who studied various aspects of the motion of these particles in the free fall in a fluid.

10.2   Proofs of Theorems 10.1 and 10.2

10.2.1   Proof of Theorem 10.1

10.2.1.1   Step 1

Decompose $b\xi(x)$ as follows:

$$(P_H b\xi)(x) = b\xi(x) - \nabla w(x), \; P_H b\xi \in H, \; b\xi \in (L^2(\Omega))^3, \; \nabla w \in G(\Omega),$$
(10.2.1)

where w solves the following generalized Neumann problem:

$$\triangle w = \text{div}\,[b\xi(x)] \quad \text{in}\;\; \Omega, \; \left.\frac{\partial w}{\partial v}\right|_{\partial\Omega} = 0.$$
(10.2.2)

Note that, since ξ vanishes in A, then $P_H b\xi = -\nabla w$ in A, and

$$\frac{1}{\text{meas}\{A\}} \int_A (Pb\xi)(x)dx = -\frac{1}{\text{meas}\{A\}} \int_A \nabla w dx.$$
(10.2.3)

Hence, to prove Theorem 10.1, it is sufficient to show that

$$\left\| \frac{1}{\text{meas}\{A\}} \int_A \nabla w dx \right\|_{R^3}$$

$$\leq \frac{C\|b\|_{R^3}}{d_*^3} \left(\text{meas}^{1/2}\{\partial\Omega\} + \text{meas}^{1/2}\{S\}\right) \text{meas}\{A\} \, \text{meas}^{1/2}\{S\}.$$
(10.2.4)

10.2.1.2   Step 2: Green's Formula

We intend to use the generalized version of the classical Green's formula representing solutions of the boundary problems (10.2.2), namely,

$$w(x) = \int_{\Omega} \Delta w(y) \, (U(x-y)) \, dy + \int_{\partial \Omega} w(\eta) \frac{\partial}{\partial \nu} (U(x-\eta)) \, d\eta, \qquad (10.2.5)$$

where

$$U(x-y) = -\frac{1}{4\pi \, \|x-y\|_{R^3}}.$$

Taking into account (10.2.2),

$$w(x) = -\frac{1}{4\pi} \int_{\partial\Omega} w(\eta) \frac{\partial}{\partial\nu} \left(\frac{1}{\|x-\eta\|_{R^3}} \right) d\eta - \frac{1}{4\pi} \int_{\Omega} \Delta w(y) \left(\frac{1}{\|x-y\|_{R^3}} \right) dy$$

$$= -\frac{1}{4\pi} \int_{\partial\Omega} w(\eta) \frac{\partial}{\partial\nu} \left(\frac{1}{\|x-\eta\|_{R^3}} \right) d\eta + \frac{1}{4\pi} \int_{S} b^T \nabla \left(\frac{1}{\|x-y\|_{R^3}} \right) dy,$$

$$\tag{10.2.6}$$

and

$$\nabla w(x) = -\frac{1}{4\pi} \nabla \int_{\partial\Omega} w(\eta) \frac{\partial}{\partial\nu} \left(\frac{1}{\|x-\eta\|_{\mathbb{R}^3}} \right) d\eta + \frac{1}{4\pi} \nabla \int_{S_\nu} b^T \nabla \left(\frac{1}{\|x-y\|_{\mathbb{R}^3}} \right) dy,$$

$$\tag{10.2.7}$$

where $y = (y_1, y_2, y_3)$ (and the second term in the first line in (10.2.6) is understood formally at this point).

Remark 10.3   Here and below, similar to the calculations in Chap. 9:

- When we write ∇ within some integral, we mean that the corresponding differentiation is conducted with respect to the integration variables.
- We interpret the improper integral over the given domain E for a function with a discontinuity at x as the limit of the integrals over $E \setminus B_s(x)$ as $s \to 0^+$.
- Note that the last integrals on the right in (10.2.6)–(10.2.7) are well defined near the "bad point" $x = (x_1, x_2, x_3)$, which can be shown by switching to the spherical coordinates near it.
- To derive (10.2.6), we can first derive it for a sequence of solutions to (10.2.2), generated by a sequence of continuously differentiable functions $g_n(x)$ on the right, which converge to $\xi(x)$ in $L^2(\Omega)$, and then pass to the limit as $n \to \infty$.

10.2.1.3   Step 3: Evaluation of the First Term on the Right in (10.2.7)
over A

Let us start with recalling the following classical embedding estimate for $H^1(\Omega)$:

$$\|\phi\|_{L^2(\partial\Omega)} \le L_o \|\phi\|_{H^1(\Omega)}, \quad \phi \in H^1(\Omega), \tag{10.2.8}$$

where L_o depends on $\partial\Omega$.

Let $\{\alpha_k\}_{k=1}^{\infty}$ ($\alpha_k \to -\infty$ as $k \to \infty$) and $\{p_k\}_{k=1}^{\infty}$ be the negative eigenvalues and orthonormalized in $L^2(\Omega)$ eigenfunctions, associated with the following spectral problem:

$$\triangle p_k = \alpha p_k, \quad \left.\frac{\partial p_k}{\partial \nu}\right|_{\partial\Omega} = 0.$$

Then, in view of (10.2.2),

$$w(x) = \sum_{k=1}^{\infty} \frac{1}{\sqrt{-\alpha_k}} \left[\int_{\Omega} \xi(x) b^T \frac{\nabla p_k}{\sqrt{-\alpha_k}} dx\right] p_k(x) + K,$$

where, without loss of generality, we can set constant $K = 0$ (since in (10.2.4) we only deal with ∇w, we are interested only in the first term here).

Hence, noticing that $\left\{\frac{\nabla p_k}{\sqrt{-\alpha_k}}\right\}_{k=1}^{\infty}$ is an orthonormal sequence in $\left(L^2(\Omega)\right)^3$, we can derive from Bessel's inequality that

$$\|w\|_{H^1(\Omega)} \le C \|b\|_{R^3} \text{meas}^{1/2}\{S\}, \tag{10.2.9}$$

where C denotes a (generic) positive constant.

10.2.1.4   Step 4

Now, for $i, j, k \in \{1, 2, 3\}$, $i \ne j$, $j \ne k$, $k \ne i$, and $x \ne y$,

$$\frac{\partial}{\partial y_i}\left(\frac{1}{\|y - x\|_{R^3}}\right) = -\frac{(y_i - x_i)}{\|y - x\|_{R^3}^3}, \tag{10.2.10}$$

$$\frac{\partial^2}{\partial x_i \partial y_i}\left(\frac{1}{\|y - x\|_{R^3}}\right) = \frac{(y_j - x_j)^2 + (y_k - x_k)^2 - 2(y_i - x_i)^2}{\|y - x\|_{R^3}^5}, \tag{10.2.11}$$

$$\frac{\partial^2}{\partial x_j \partial y_i}\left(\frac{1}{\|y - x\|_{R^3}}\right) = -\frac{3(y_i - x_i)(y_j - x_j)}{\|y - x\|_{R^3}^5}. \tag{10.2.12}$$

Let d_α be the distance between the sets A and $\partial\Omega$:

$$d_\alpha = \inf_{x \in A, \eta \in \partial\Omega} \|x - \eta\|_{R^3}. \tag{10.2.13}$$

10.2.1.5   Step 5

Let $v = (v_1, v_2, v_3)$, $\|v\|_{\mathbb{R}^3} = 1$.

Remark 10.4 Making use of long but rather straightforward calculations, we can derive from (10.2.8)–(10.2.13) the following estimates:

- $$\left\| \int_A \nabla \int_{\partial\Omega} \left[w(\eta) \frac{\partial}{\partial v} \left(\frac{1}{\|x - \eta\|_{R^3}} \right) \right] d\eta dx \right\|_{R^3}$$

$$\leq \frac{L_0 C \, \|b\|_{R^3}}{d_\alpha^3} \, \mathrm{meas}^{1/2}\{\partial\Omega\} \, \mathrm{meas}\{A\} \mathrm{meas}^{1/2}\{S\}, \qquad (10.2.14)$$

where C is a (generically denoted) constant.

- $$\left\| \int_S \nabla \int_{\partial\Omega} \left[w(\eta) \frac{\partial}{\partial v} \left(\frac{1}{\|x - \eta\|_{R^3}} \right) \right] d\eta dx \right\|_{R^3}$$

$$\leq \frac{L_0 C \, \|b\|_{R^3}}{d_{sp}^3} \mathrm{meas}^{1/2}\{\partial\Omega\} \, \mathrm{meas}^{3/2}\{S\}, \qquad (10.2.15)$$

where d_{sp} denotes the distance between the sets S and $\partial\Omega$. Respectively, we have:

$$\int_S \nabla \int_{\partial\Omega} \left[w(\eta) \frac{\partial}{\partial v} \left(\frac{1}{\|x - \eta\|_{R^3}} \right) \right] d\eta dx = O(p) \, \|b\|_{R^3} \quad \text{as} \quad p \to 0^+,$$

$$(10.2.16)$$

where p is the largest dimension of parallelepiped S.
- Denote

$$d_\beta = \inf_{x \in S_o, y \in Q} \|y - x\|_{R^3} .$$

Then,

$$\left\| \int_A \nabla \int_S b^T \nabla \left(\frac{1}{\|x - y\|_{R^3}} \right) dy dx \right\|_{R^3} \leq \frac{C \, \|b\|_{R^3}}{d_\beta^3} \mathrm{meas}\{A\} \, \mathrm{meas}\{S\},$$

$$(10.2.17)$$

where C is a (generic) positive constant.

Combining (10.2.14) with (10.2.17) yields (10.2.4), which provides the result of Theorem 10.1.

10.2.2   Proof of Theorem 10.2

10.2.2.1   Step 1

Due to (10.2.1), similar to (10.2.3), we have

$$\frac{1}{\text{meas}\{S\}} \int_S (P_H b\xi)(x)dx = b - \frac{1}{\text{meas}\{S\}} \int_S \nabla w \, dx. \qquad (10.2.18)$$

In turn, (10.2.16) and (10.2.7) provide the last term in (10.1.2).
It remains to evaluate the integral of the second term in (10.2.7) over S.

10.2.2.2   Step 2

For an arbitrary $x = (x_1, x_2, x_3) \in S$, consider an $h = h(x) > 0$ such that $B_{h(x)}(x) \subset S$ and $B_{h(x)}(x)$ is strictly separated from ∂S_o.
Then,

$$\int_S b^T \nabla \frac{1}{\|x - y\|_{R^3}} dy = \int_{S \setminus B_{h(x)}(x)} b^T \nabla \frac{1}{\|x - y\|_{R^3}} dy + \int_{B_{h(x)}(x)} b^T \nabla \frac{1}{\|x - y\|_{R^3}} dy.$$
$$(10.2.19)$$

Making use of (10.2.10), we obtain

$$\int_{B_{h(x)}(x)} \frac{\partial}{\partial y_1} \left(\frac{1}{\|x - y\|_{R^3}} \right) dy = \lim_{s \to 0^+} \int_{B_{h(x)}(x) \setminus B_s(x)} \frac{x_1 - y_1}{\|x - y\|_{R^3}^3} dy, \qquad (10.2.20)$$

where $B_s(x)$ is the ball with center at x of radius s, $0 < s < h(x)$.
Recalling the standard spherical coordinates:

$$y_1 - x_1 = r \cos \alpha, \qquad s \le r \le h,$$
$$y_2 - x_2 = r \sin \alpha \cos \beta, \quad 0 \le \alpha \le \pi, \quad dy = r^2 \sin \alpha \, dr d\alpha d\beta,$$
$$y_3 - x_3 = r \sin \alpha \sin \beta, \quad 0 \le \beta \le 2\pi,$$

we can derive that

$$\int_{B_{h(x)}(x) \setminus B_s(x)} \frac{x_1 - y_1}{\|x - y\|_{R^3}^3} dy = -\frac{1}{2} \int_0^{2\pi} d\beta \int_0^\pi \sin 2\alpha \, d\alpha \int_s^{h(x)} dr = 0.$$
$$(10.2.21)$$

It follows from (10.2.20)–(10.2.21) that

$$\int_{B_{h(x)}(x)} \frac{\partial}{\partial y_1} \left(\frac{1}{\|x - y\|_{R^3}} \right) dy = 0. \qquad (10.2.22)$$

Similar calculations for the integrations with respect to y_2 and y_3 yield

$$\int_{B_{h(x)}(x)} b^T \nabla \frac{1}{\|x - y\|_{\mathbb{R}^3}} dy = 0. \tag{10.2.23}$$

Thus, making use of (10.2.19) and (10.2.23), we can derive that

$$\int_S b^T \nabla \frac{1}{\|x - y\|_{R^3}} dy = \int_{S \setminus B_{h(x)}(x)} \frac{b_1(x_1 - y_1) + b_2(x_2 - y_2) + b_3(x_3 - y_3)}{\|x - y\|_{R^3}^3} dy. \tag{10.2.24}$$

10.2.2.3 Step 3

We intend now to evaluate the second term in (10.2.7).

Fix any $x = (x_1, x_2, x_3) \in S$. Due to our selection of h for a given x, for small $\triangle x_1$, there exists

$$x_{\delta 1} = (x_1 + \triangle x_1, x_2, x_3), \quad h(x_{\delta 1}) > h(x),$$

such that

$$B_{h(x)}(x_{\delta 1}) \subset B_{h(x_{\delta 1})}(x_{\delta 1}) \subset S.$$

Note now that, as in (10.2.21), we have

$$\int_{B_{h(x_{\delta 1})}(x_{\delta 1}) \setminus B_{h(x)}(x_{\delta 1})} \frac{\partial}{\partial y_1} \left(\frac{1}{\|x_{\delta 1} - y\|_{\mathbb{R}^3}} \right) dy = 0. \tag{10.2.25}$$

From here, we obtain, via (10.2.24) and (10.2.25), that

$$\frac{\partial}{\partial x_1} \left(\int_S b^T \nabla \frac{1}{\|x - y\|_{R^3}} dy \right)$$

$$= \lim_{\triangle x_1 \to 0} \frac{1}{\triangle x_1} \left(\int_{S \setminus B_{h(x_{\delta 1})}(x_{\delta 1})} b^T \nabla \frac{1}{\|x_{\delta 1} - y\|_{R^3}} dy - \int_{S \setminus B_{h(x)}(x)} b^T \nabla \frac{1}{\|x - y\|_{R^3}} dy \right)$$

$$= \lim_{\triangle x_1 \to 0} \frac{1}{\triangle x_1} \left(\int_{S \setminus B_{h(x)}(x_{\delta 1})} b^T \nabla \frac{1}{\|x_{\delta 1} - y\|_{R^3}} dy - \int_{S \setminus B_{h(x)}(x)} b^T \nabla \frac{1}{\|x - y\|_{R^3}} dy \right)$$

$$= \sum_{i=1}^{3} b_i \frac{\partial}{\partial x_1} \left(\int_{S \setminus B_h(x)} \frac{x_i - y_i}{\|x - y\|_{R^3}^3} dy \right), \tag{10.2.26}$$

where $h = h(x)$ on the last line in (10.2.26) is now treated as independent of x when calculating the derivatives.

10.2.2.4   Step 4: Calculation of the Terms in the Last Line in (10.2.26)

To simplify notations, we will further write h instead of $h(x)$.

Remark 10.5 Once again, making use of long but rather straightforward calculations (see [] for all the details), we can derive the following equalities:

-

$$\frac{\partial}{\partial x_1} \left(\int_{S \backslash B_h(x)} \frac{x_1 - y_1}{\|x - y\|_{R^3}^3} dy \right) = \frac{4\pi}{3} + \int_{S \backslash B_h(x)} \frac{\partial}{\partial x_1} \left(\frac{x_1 - y_1}{\|x - y\|_{R^3}^3} \right) dy,$$

(10.2.27)

-

$$\frac{\partial}{\partial x_2} \left(\int_{S \backslash B_h(x)} \frac{x_2 - y_2}{\|x - y\|_{R^3}^3} dy \right) = \frac{4\pi}{3} + \int_{S \backslash B_h(x)} \frac{\partial}{\partial x_2} \left(\frac{x_2 - y_2}{\|x - y\|_{R^3}^3} \right) dy,$$

(10.2.28)

-

$$\frac{\partial}{\partial x_3} \left(\int_{S \backslash B_h(x)} \frac{x_3 - y_3}{\|x - y\|_{R^3}^3} dy \right) = \frac{4\pi}{3} + \int_{S \backslash B_h(x)} \frac{\partial}{\partial x_3} \left(\frac{x_3 - y_3}{\|x - y\|_{R^3}^3} \right) dy.$$

(10.2.29)

-

$$\frac{\partial}{\partial x_j} \left(\int_{S \backslash B_h(x)} \frac{x_i - y_i}{\|x - y\|_{R^3}^3} dy \right) = \int_{S \backslash B_h(x)} \frac{\partial}{\partial x_j} \left(\frac{x_i - y_i}{\|x - y\|_{R^3}^3} \right) dy, \ \ i, j \in \{1, 2, 3\}, \ i \neq j.$$

(10.2.30)

Formulas (10.2.27)–(10.2.30) provide the formula for g in (10.1.3)–(10.1.5) (see Step 1 of the proof).

This completes the proof of Theorem 10.2.

10.3   Proofs of Main Results

10.3.1   Proofs of Theorems 10.3–10.5

10.3.1.1   Auxiliary Formulas

Due to (10.1.2), to establish (10.1.7), we need to evaluate the integrals in (10.1.3)–(10.1.5).

Select $B_h(x)$ as follows:

$$B_{h(x)}(x) = \left\{ y : \|y - x\|_{R^3} < h(x) = \frac{1}{2} \min \{ p - |x_1| \, ; \, q - |x_2| \, ; \, \ell - |x_3| \} \right\} \subset S.$$

Due to the antisymmetry of the linear functions and our "symmetric" choice of $h(x)$, i.e.,

$$h(x_1, x_2, x_3) = h(-x_1, x_2, \ x_3) = h(-x_1, -x_2, \ x_3) = h(-x_1, -x_2, -x_3)$$

$$= h(x_1, \ -x_2, -x_3) = h(x_1, x_2, \ -x_3) = h(-x_1, x_2, \ -x_3) = h(x_1, -x_2, x_3),$$

we have

$$\frac{1}{\text{meas}\{S\}} \int_S \int_{S \setminus B_h(x)} \frac{\partial}{\partial x_j} \left(\frac{x_i - y_i}{\|x - y\|_{R^3}^3} \right) dy dx = 0, \quad i, j \in \{1, 2, 3\}, \ i \neq j.$$

$$(10.3.1)$$

To simplify further notation, set

$$\begin{aligned} p_0 &= p + x_1 > 0, \quad q_0 = q + x_2 > 0, \quad \ell_0 = \ell + x_3 > 0, \\ p_1 &= p - x_1 > 0, \quad q_1 = q - x_2 > 0, \quad \ell_1 = \ell - x_3 > 0, \end{aligned} \qquad (10.3.2)$$

$$P_i = \int_{S \setminus B_h(x)} \frac{\partial}{\partial x_i} \left(\frac{x_i - y_i}{\|x - y\|_{R^3}^3} \right) dy. \qquad (10.3.3)$$

Remark 10.6 Once again, making use of long but rather straightforward calculations (see [] for all the details), we can derive the following equalities:

-
$$\frac{1}{\text{mes}\{S\}} \int_S P_1 dx = -\frac{4\pi}{3} + \frac{8}{\text{meas}\{S\}} \int_S \tan^{-1} \frac{q_1 \ell_1}{p_1 \sqrt{p_1^2 + q_1^2 + \ell_1^2}} dx,$$

$$(10.3.4)$$

-
$$\frac{1}{\text{meas}\{S\}} \int_S P_2 dx = -\frac{4\pi}{3} + \frac{8}{\text{meas}\{S\}} \int_S \tan^{-1} \frac{p_1 \ell_1}{q_1 \sqrt{p_1^2 + q_1^2 + \ell_1^2}} dx,$$

$$(10.3.5)$$

-
$$\frac{1}{\text{meas}\{S\}} \int_S P_3 dx = -\frac{4\pi}{3} + \frac{8}{\text{meas}\{S\}} \int_S \tan^{-1} \frac{q_1 p_1}{\ell_1 \sqrt{p_1^2 + q_1^2 + \ell_1^2}} dx.$$

$$(10.3.6)$$

In view of (10.2.10)–(10.2.11), we have

$$\frac{\partial}{\partial x_1}\left(\frac{x_1 - y_1}{\|x - y\|_{R^3}^3}\right) + \frac{\partial}{\partial x_2}\left(\frac{x_2 - y_2}{\|x - y\|_{R^3}^3}\right) + \frac{\partial}{\partial x_3}\left(\frac{x_3 - y_3}{\|x - y\|_{R^3}^3}\right) = 0. \quad (10.3.7)$$

Combining (10.3.1), (10.3.3), and (10.3.7), we obtain

$$\frac{1}{\text{meas}\{S\}} \int_S P_1 dx + \frac{1}{\text{meas}\{S\}} \int_S P_2 dx + \frac{1}{\text{meas}\{S\}} \int_S P_3 dx = 0. \quad (10.3.8)$$

10.3.1.2   Proof of Theorem 10.3

Note the assumption of Theorem 2.1 that $p \geq q$ also implies that

$$\ell^{1-\varepsilon}/p \to 0 \text{ as } p \to 0.$$

Introduce the following set:

$$A(p, q, \ell) := \left\{(x_1, x_2, x_3) : |x_1| < p - \ell^{1-\varepsilon}, |x_2| < q - \ell^{1-\varepsilon}, |x_3| < \ell\right\}. \quad (10.3.9)$$

Without loss of generality, we can assume that

$$A(p, q, \ell) \subset S.$$

Due to (10.3.9), we have

$$0 < \frac{\ell_1}{p_1} = \frac{\ell - x_3}{p - x_1} < \frac{2\ell}{\ell^{1-\varepsilon}} = 2\ell^\varepsilon \ \ \forall (x_1, x_2, x_3) \in A(p, q, \ell), \quad (10.3.10)$$

$$0 < \frac{\ell_1}{q_1} = \frac{\ell - x_3}{q - x_2} < \frac{2\ell}{\ell^{1-\varepsilon}} = 2\ell^\varepsilon \ \ \forall (x_1, x_2, x_3) \in A(p, q, \ell). \quad (10.3.11)$$

Step 1: Evaluation of $\frac{1}{\text{meas}\{S\}} \int_S P_1 dx$

Making use of (10.3.10) and monotonicity of the $\tan^{-1}$-function, we obtain that

$$0 < \tan^{-1} \frac{q_1 \ell_1}{p_1 \sqrt{p_1^2 + q_1^2 + \ell_1^2}} < \tan^{-1} \frac{\ell_1}{p_1} < \tan^{-1} 2\ell^\varepsilon < 2\ell^\varepsilon.$$

This will give us the following chain of estimates:

$$0 < \frac{1}{\text{meas}\{S\}} \int_{A(p,q,\ell)} \tan^{-1} \frac{q_1\ell_1}{p_1\sqrt{p_1^2 + q_1^2 + \ell_1^2}} dx < \frac{\text{meas}\{A(p,q,\ell)\}}{\text{meas}\{S\}} 2\ell^\varepsilon < 2\ell^\varepsilon.$$

(10.3.12)

Since $0 < \tan^{-1} s < \frac{\pi}{2} \ \forall s > 0$, we have

$$0 < \frac{1}{\text{meas}\{S\}} \int_{S\backslash A(p,q,\ell)} \tan^{-1} \frac{q_1\ell_1}{p_1\sqrt{p_1^2 + q_1^2 + \ell_1^2}} dx < \frac{\text{meas}\{S \backslash A(p,q,\ell)\}}{\text{meas}\{S\}} \frac{\pi}{2},$$

(10.3.13)

where

$$\frac{\text{meas}\{S \backslash A(p,q,\ell)\}}{\text{meas}\{S\}} = 1 - (1 - \frac{\ell^{1-\varepsilon}}{q})^2 < 2\frac{\ell^{1-\varepsilon}}{q}.$$

Thus,

$$0 < \frac{1}{\text{meas}\{S\}} \int_{S\backslash A(p,q,\ell)} \tan^{-1} \frac{q_1\ell_1}{p_1\sqrt{p_1^2 + q_1^2 + \ell_1^2}} dx < \pi \frac{\ell^{1-\varepsilon}}{q}. \quad (10.3.14)$$

Combining (10.3.14) with (10.3.12) will give us

$$0 < \frac{1}{\text{meas}\{S\}} \int_{S} \tan^{-1} \frac{q_1\ell_1}{p_1\sqrt{p_1^2 + q_1^2 + \ell_1^2}} dx < 2\ell^\varepsilon + +\pi \frac{\ell^{1-\varepsilon}}{q}.$$

Hence,

$$\frac{1}{\text{meas}\{S\}} \int_{S} P_1 dx = -\frac{4\pi}{3} + O(\ell^\varepsilon) + O(\frac{\ell^{1-\varepsilon}}{q}). \quad (10.3.15)$$

Step 2: Evaluation of $\frac{1}{\text{meas}\{S\}} \int_S P_2 dx$

Employing (10.3.11) for

$$(x_1, x_2, x_3) \subset \Lambda(p, q, \ell),$$

similarly to (10.3.12) and (10.3.14), we can derive

$$\frac{1}{\text{meas}\{S\}} \int_{S} P_2 dx = -\frac{4\pi}{3} + O(\ell^\varepsilon) + O(\frac{\ell^{1-\varepsilon}}{q}). \quad (10.3.16)$$

Step 3: Evaluation of $\frac{1}{\text{meas}\{S\}} \int_S P_3 dx$

We can combine (10.3.16), (10.3.15), and (10.3.8) to obtain

$$\frac{1}{\text{meas}\{S\}} \int_S P_3 dx = \frac{8\pi}{3} + O(\ell^\varepsilon) + O(\frac{\ell^{1-\varepsilon}}{q}). \qquad (10.3.17)$$

Step 4
Combining (10.3.15), (10.3.16), and (10.3.17) yields that in (10.1.2)–(10.1.5) for our S, we have

$$\frac{1}{\text{meas}\{S\}} \int_S g_1 dx = \left[O(\ell^\varepsilon) + O(\frac{\ell^{1-\varepsilon}}{q}) + O(p) \right] \|b\|_{R^3}, \qquad (10.3.18)$$

$$\frac{1}{\text{meas}\{S\}} \int_S g_2 dx = \left[O(\ell^\varepsilon) + O(\frac{\ell^{1-\varepsilon}}{q}) + O(p) \right] \|b\|_{R^3}, \qquad (10.3.19)$$

$$\frac{1}{\text{meas}\{S\}} \int_S g_3 dx = b_3 + \left[O(\ell^\varepsilon) + O(\frac{\ell^{1-\varepsilon}}{q}) + O(p) \right] \|b\|_{R^3}. \quad (10.3.20)$$

Combining (10.3.18)–(10.3.20) and (10.3.1) with (10.1.2) will give us (10.1.7), which completes the proof of Theorem 10.3.

10.3.1.3   Proof of Theorem 10.4

Set

$$B(p, q, q) := \left\{ (x_1, x_2, x_3) : |x_1| < p - q^{1-\tau}, |x_2| < q, |x_3| < q \right\}, \qquad (10.3.21)$$

$$\text{meas}\{B(p, q, q)\} = 8pqq \left(1 - \frac{q^{1-\tau}}{p} \right). \qquad (10.3.22)$$

As $p \to 0$, under the assumptions of Theorem 10.4, without loss of generality, we can assume that

$$B(p, q, q) \subset S.$$

In turn, due to (10.3.21), we have

$$0 < \frac{\ell_1}{p_1} = \frac{q - x_3}{p - x_1} < \frac{2q}{q^{1-\tau}} = 2q^\tau \quad \forall (x_1, x_2, x_3) \in B(p, q, q). \; (10.3.23)$$

Step 1: Evaluation of $\frac{1}{\text{meas}\{S\}} \int_S P_1 dx$
As in Step 1 in the proof of Theorem 10.3, we can obtain

$$0 < \tan^{-1} \frac{q_1 \ell_1}{p_1 \sqrt{p_1^2 + q_1^2 + \ell_1^2}} < 2q^\tau,$$

$$0 < \frac{1}{\text{meas}\{S\}} \int_{B(p,q,q)} \tan^{-1} \frac{q_1 \ell_1}{p_1 \sqrt{p_1^2 + q_1^2 + \ell_1^2}} dx < \frac{\text{meas}\{B(p,q,q)\}}{\text{meas}\{S_o\}} 2q^\tau < 2q^\tau,$$

$$(10.3.24)$$

$$0 < \frac{1}{\text{meas}\{S\}} \int_{S \setminus B(p,q,q)} \tan^{-1} \frac{q_1 \ell_1}{p_1 \sqrt{p_1^2 + q_1^2 + \ell_1^2}} dx < \frac{\text{meas}\{S \setminus B(p,q,q)\}}{\text{meas}\{S\}} \frac{\pi}{2},$$

where

$$\frac{\text{meas}\{S \setminus B(p,q,q)\}}{\text{meas}\{S\}} = 1 - \left(1 - \frac{q^{1-\tau}}{p}\right) = \frac{q^{1-\tau}}{p}.$$

Thus,

$$0 < \frac{1}{\text{meas}\{S\}} \int_{S \setminus B(p,q,q)} \tan^{-1} \frac{q_1 \ell_1}{p_1 \sqrt{p_1^2 + q_1^2 + \ell_1^2}} dx < \frac{\pi}{2} \frac{q^{1-\tau}}{p} \quad \text{as } p \to 0.$$

$$(10.3.25)$$

Combining (10.3.24) and (10.3.25) provides us with

$$0 < \frac{1}{\text{meas}\{S\}} \int_S \tan^{-1} \frac{q_1 \ell_1}{p_1 \sqrt{p_1^2 + q_1^2 + \ell_1^2}} dx < 2q^\tau + \frac{\pi}{2} \frac{q^{1-\tau}}{p}.$$

Hence,

$$\frac{1}{\text{meas}\{S\}} \int_S P_1 dx = -\frac{4\pi}{3} + O(q^\tau) + O(\frac{q^{1-\tau}}{p}). \qquad (10.3.26)$$

Step 2: Evaluation of $\frac{1}{\text{meas}\{S_o\}} \int_S P_2 dx$ **and** $\frac{1}{\text{meas}\{S\}} \int_S P_3 dx$

By switching the role of variables x_2 and x_3, we get

$$\int_{-p}^{p} \int_{-q}^{q} \int_{-q}^{q} \tan^{-1} \frac{(p - x_1)(q - x_3)}{(q - x_2)\sqrt{(p - x_1)^2 + (q - x_2)^2 + (q - x_3)^2}} dx_3 dx_2 dx_1$$

$$= \int_{-p}^{p} \int_{-q}^{q} \int_{-q}^{q} \tan^{-1} \frac{(p - x_1)(q - x_2)}{(q - x_3)\sqrt{(p - x_1)^2 + (q - x_2)^2 + (q - x_3)^2}} dx_3 dx_2 dx_1.$$

Thus,

$$\frac{1}{\text{meas}\{S\}}\int_S \tan^{-1}\frac{p_1\ell_1}{q_1\sqrt{p_1^2+q_1^2+\ell_1^2}}dx = \frac{1}{\text{meas}\{S\}}\int_S \tan^{-1}\frac{p_1 q_1}{\ell_1\sqrt{p_1^2+q_1^2+\ell_1^2}}dx.$$

$$(10.3.27)$$

Now, we will combine (10.3.27), (10.3.6), (10.3.5), (10.3.8), and (10.3.26) to derive

$$\frac{1}{\text{meas}\{S\}}\int_S P_2 dx = \frac{1}{\text{meas}\{S\}}\int_S P_3 dx = \frac{2\pi}{3}+O(q^\tau)+O(\frac{q^{1-\tau}}{p}). \quad (10.3.28)$$

Step 3

Combining (10.3.26) and (10.3.28), we can show that in (10.1.3)–(10.1.5) for our S we have

$$\frac{1}{\text{meas}\{S\}}\int_S g_1 dx = \left[O(q^\tau)+O(\frac{q^{1-\tau}}{p})+O(p)\right]\|b\|_{R^3}, \quad (10.3.29)$$

$$\frac{1}{\text{meas}\{S\}}\int_S g_2 dx = \frac{b_2}{2}+\left[O(q^\tau)+O(\frac{q^{1-\tau}}{p})+O(p)\right]\|b\|_{R^3}, \quad (10.3.30)$$

$$\frac{1}{\text{meas}\{S\}}\int_S g_3 dx = \frac{b_3}{2}+\left[O(q^\tau)+O(\frac{q^{1-\tau}}{p})+O(p)\right]\|b\|_{R^3}. \quad (10.3.31)$$

If we combine (10.3.29)–(10.3.31) and (10.3.1) with (10.1.2), we can derive (10.1.8), which completes the proof of Theorem 10.4.

10.3.1.4   Proof of Theorem 10.5

It follows immediately from Theorem 10.2, since, due to the symmetry of a ball, the integrals of all the integrals in (10.1.3)–(10.1.5) (needed in (10.1.2)) are all equal to zero.

Remark 10.7 (Open Questions and Research Perspectives) In this chapter, the general formula for

$$\frac{1}{\text{meas}\{S\}}\int_S (P_H b\xi)(x)dx$$

in Theorem 10.2 was used to obtain specific results for the case when S is a parallelepiped. Based on the aforementioned general theorem, one can similarly investigate other geometric shapes of S (e.g., ellipses) to be used in the design of various bio-mimetic swimmers, or even the sets S that represent the geometric shapes of various living organisms, see also Remarks 9.4, 11.3.

Part V
Global Steering for Bio-Mimetic Swimmers in 2D and 3D Incompressible Fluids

Chapter 11
Swimming Capabilities of Swimmers in 2D and 3D Incompressible Fluids: Force Controllability

In this chapter we will apply the results of Chaps. 7–10 to propose *a strategy of designing bio-mimetic swimmers* which, theoretically, can make use of their internal forces and geometric controls, to propel themselves from any initial position to any desirable position within a fluid domain, while preserving the structural integrity of their bodies as stated in Sect. 5.2 in Chap. 5.

11.1 Discussion of Concepts for Global Swimming Locomotion

Our main idea here comes from the *swimming strategy applied by the living organisms and existing artificial devices* that are observed in the real world; namely,

1. each of such organisms and devices has *limited capabilities* in terms of creating a force that can be engaged to ensure their momentary desirable swimming locomotion;
2. a typical strategy of the real-world organisms and devices consists of *applying of a force that will momentarily propel them in the desirable direction*;
3. in terms of reaching a desirable position in space, *the success of such strategy is not always guaranteed* and depends heavily on the evolution of velocity of the surrounding medium.

For example, a real-world swimmer cannot successfully swim against a sufficiently strong current or maintain a desirable direction for motion under the conditions of strong turbulence.

Remark 11.1 (Comments on a Mathematical Controllability) A typical mathematical controllability theory does not assume any restrictions on the size of allowed controls, which are usually elements of a vector (normed) space. This includes our

© The Author(s), under exclusive license to Springer Nature Switzerland AG 2021
A. Khapalov, *Bio-Mimetic Swimmers in Incompressible Fluids*, Lecture Notes
in Mathematical Fluid Mechanics, https://doi.org/10.1007/978-3-030-85285-6_11

results on local controllability in Chaps. 7 and 8 (and on global controllability in [25], Ch. 15). However, in these chapters, we also derived the formulas (7.2.8) and (8.2.1) for swimmers' micromotions that allow us to evaluate the requirements on control forces needed to achieve a desirable momentary velocity, both in magnitude and in direction.

We will now introduce two definitions of controllability which can characterize swimming capabilities of a swimmer at hand.

Definition 11.1 We will say that a swimmer is **force controllable**, if, regardless of its immediate skeleton shape (i.e., positions of z_i's in the fluid domain), the set of all possible forces to act on the swimmer's center of mass in an incompressible fluid that its internal forces and geometric controls can create contains a neighborhood of the origin in the respective space R^K, $K = 2, 3$.

This definition means that a force controllable swimmer can momentarily move its center of mass at any desirable direction.

Definition 11.2 Given an immediate skeleton shape of a swimmer (i.e., positions of z_i's in the fluid domain) and a vector **a**, we will say that this swimmer is **a-directionally force controllable from its immediate skeleton position** if the set of all possible forces to act on the swimmer's center of mass in an incompressible fluid that its internal forces and geometric controls can create for the given skeleton position contains a force that will point at the direction co-linear with **a**.

This definition means that an **a**-directionally force controllable swimmer can momentarily move its center of mass at the direction co-linear with **a**.

Remark 11.2 It should be noted that the actual patterns of swimming locomotion may include both "useful" motion phases (the locomotion in a desirable direction) and various auxiliary phases (e.g., a recovery phase in a breaststroke swimming locomotion, see Sect. 11.3.3 below for illustration).

11.2   An Instrumental Observation

Below we will consider the swimmers discussed in Chaps. 7–10 that can engage the rotational forces and controlled elastic forces only (no Hooke's elastic forces, for simplicity of further discussion).

Making use of the formulas (8.1.6)–(8.1.7), we can represent a force term for such swimmers as follows:

$$F = \sum_{i=1}^{2n-3} v_i F_i, \tag{11.2.1}$$

where

$$v = (v_1, \ldots, v_{n-2}, w_1, \ldots, w_{n-1}) = (v_1, \ldots, v_{2n-3}), \qquad (11.2.2)$$

where $v_i, i = 1, \ldots, n - 2$, and $v_{n-1} = w_1, \ldots, v_{2n-3} = w_{n-1}$ are selectable controls as in Assumptions 7.2 and 8.1.

Observe the following:

- Each F_i is an element of $(L^\infty(\Omega))^K$, $K = 2, 3$, with support restricted to finitely many sets among all swimmer's body parts $S(z_j(t))$'s.
- The projections $P_H F_i$'s of these forces on the fluid velocity space H are the actual forces that act on a swimmer in a fluid. We studied them in Chaps. 9 and 10.
- More precisely, for small swimmers, we showed in Chaps. 9 and 10 that only the averaged values of $P_H F_i$'s over the original supports of F_i's are important, while their actions on other body parts of a small swimmer are negligible, see Theorems 9.1 and 10.1.

11.3 Illustrating Examples in 2D: A Snake- or Fish-Like and Breaststroke Locomotions

In this section we will consider several illustrating examples, assuming that our swimmer consists of small narrow rectangles, and, according to Theorem 9.3, any force applied to such a rectangle will retain, *in an incompressible fluid*, only a component that will be its projection on the longer dominating dimension of the respective rectangle.

We consider a swimmer that has only three body parts. However, this swimmer can be viewed as a part of a larger swimmer.

11.3.1 Fish- or Snake-Like Locomotion to the Left

- In Figs. 11.1, 11.2, and 11.3, we show possible sets of internal rotational and elastic forces and geometric controls, engaged by our swimmer outside an incompressible fluid.
- For each of the above figures, in Figs. 11.4, 11.5, and 11.6, we show the actual forces that will act on each rectangle inside an incompressible fluid.
- The swimmer moves to the left, but each rectangle will also move a little up or down in a snake- or fish-like motion.

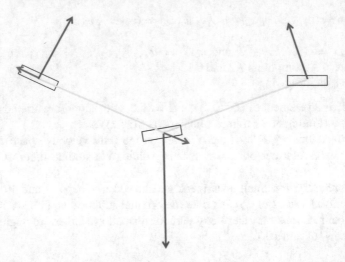

Fig. 11.1 Fish- or snake-like locomotion: forces acting outside of an incompressible medium

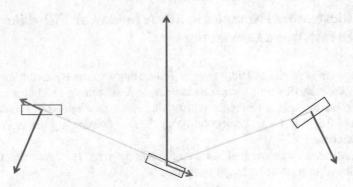

Fig. 11.2 Fish- or snake-like locomotion: forces acting outside of an incompressible medium

11.3.2   Turning Motion of One Rectangle, While the Other Two Retain Their Position

It is shown in Figs. 11.7 and 11.8.

11.3.3   Breaststroke Locomotion for a Swimmer Consisting of 3 Rectangles: A Bio-Mimetic Clam (Scallop)

• In Fig. 11.9, we show possible sets of internal rotational and elastic forces and geometric controls, engaged by our swimmer outside an incompressible fluid *in the "useful motion" phase*.

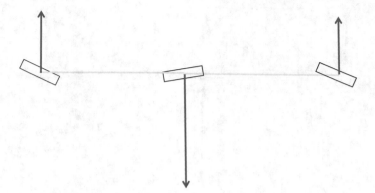

Fig. 11.3 Fish- or snake-like locomotion: forces acting outside of an incompressible medium

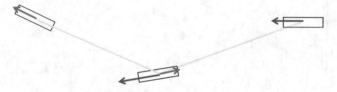

Fig. 11.4 Fish- or snake-like locomotion: transformed forces from Fig. 11.1 acting inside of an incompressible medium

Fig. 11.5 Fish- or snake-like locomotion: transformed forces from Fig. 11.2 acting inside of an incompressible medium

Fig. 11.6 Fish- or snake-like locomotion: transformed forces from Fig. 11.3 acting inside of an incompressible medium

- In Fig. 11.10 we show the actual forces that will act on each rectangle inside an incompressible fluid *in the "useful motion" phase*.
- The swimmer's center of mass moves up *in the "useful motion" phase*.
- In Fig. 11.11, we show possible sets of internal rotational and elastic forces and geometric controls, engaged by our swimmer outside an incompressible fluid *in the "recovery" phase*.

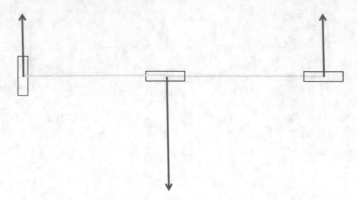

Fig. 11.7 Turning locomotion: forces acting outside of an incompressible medium

Fig. 11.8 Turning locomotion: transformed forces from Fig. 11.7 acting inside of an incompressible medium

Fig. 11.9 Breaststroke locomotion for a bio-mimetic clam in the "useful motion" phase: forces acting outside of an incompressible medium

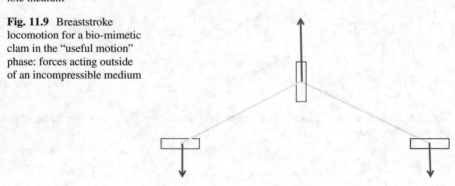

- In Fig. 11.12, we show the actual forces that will act on each rectangle inside an incompressible fluid *in the "recovery" phase*.
- The swimmer's center of mass moves up *in the "recovery" phase* as well.

Fig. 11.10 Breaststroke
locomotion for a bio-mimetic
clam in the "useful motion"
phase: transformed forces
from Fig. 11.9 acting inside
of an incompressible medium

Fig. 11.11 Breaststroke
locomotion for a bio-mimetic
clam in the "recovery" phase:
"recharging" forces acting
outside of an incompressible
medium

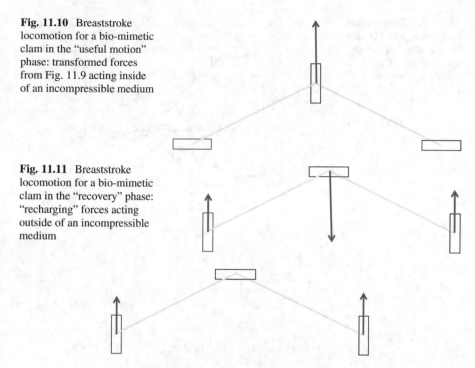

Fig. 11.12 Breaststroke locomotion for a bio-mimetic clam in the "recovery" phase: "recharging" transformed forces from Fig. 11.11 acting inside of an incompressible medium

11.3.4 Breaststroke Locomotion for a Swimmer Consisting of 5 Rectangles: A Bio-Mimetic Aquatic Frog

- In Fig. 11.13, we show possible sets of internal rotational and elastic forces and geometric controls, engaged by our swimmer outside an incompressible fluid.
- In Fig. 11.14, we show the actual forces that will act on each rectangle inside an incompressible fluid.
- The swimmer's center of mass moves to the left.

11.4 Breaststroke Pattern for a Swimmer Consisting of 3 Discs

- In Fig. 11.15, we show the same internal rotational and elastic forces and geometric controls as on Fig. 11.9 (i.e., for a swimmer consisting of 3 rectangles), engaged by our swimmer outside an incompressible fluid.

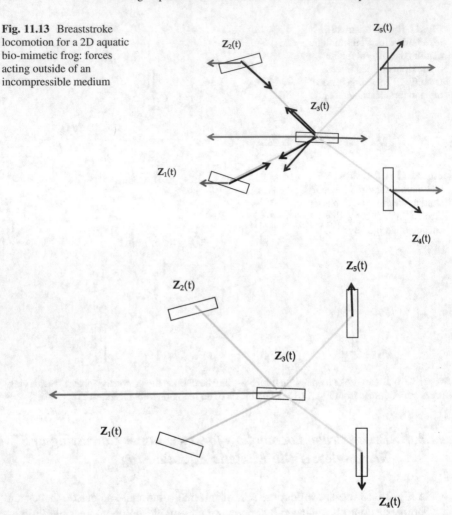

Fig. 11.13 Breaststroke locomotion for a 2D aquatic bio-mimetic frog: forces acting outside of an incompressible medium

Fig. 11.14 Breaststroke locomotion for a 2D aquatic bio-mimetic frog: transformed forces from Fig. 11.13 acting inside of an incompressible medium

- In Fig. 11.16, we show the actual forces that will act on each disc inside an incompressible fluid: all forces retain the same directions but are reduced in their magnitude by the same factor, as stated in Theorem 9.4.
- The swimmer's center of mass does not move under the above forces: the sum of all forces acting on the center of mass remains to be zero.

Fig. 11.15 Breaststroke
pattern for a swimmer
consisting of 3 discs: forces
acting outside of an
incompressible medium

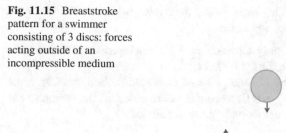
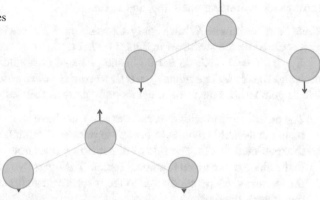

Fig. 11.16 Breaststroke pattern for a swimmer consisting of 3 discs: transformed forces from Fig. 11.15 acting inside of an incompressible medium. The sum of all forces remains to be zero: no locomotion

11.5 Illustrating Examples in 3D

We cannot provide many illustrating examples in 3D, because it is difficult to illustrate a 3D motion with forces, acting in 3D, in the context of the two-dimensional format of this monograph. Let us only point out at the following particular simple situations.

Consider a swimmer that consists of 3 small identical parallelepipeds with dimensions $\ell << q \leq p$ as in Theorem 10.3.

Case 1: A Swimmer Whose Body Consists of 3 Parallelepipeds: Locomotion in One Plane

- Assume that all three centers of mass of the swimmer's body parts lie in the plane that contains a desirable momentary motion for the swimmer.
- Then, the swimmer can apply the geometric controls to orient all its body parts in such a way that the intermediate sides of length q of each body part will be perpendicular to the aforementioned plane for a desirable momentary motion.
- After that, all the internal forces that the swimmer can engage will be lying in the plane of the desirable momentary motion, and Theorem 10.3 yields that, in the fluid, these forces will retain only the components that are their projections on the longest sides of length p.
- Thus, the momentary steering problem for this swimmer becomes two dimensional and all the strategies from Sect. 11.3 apply.

Case 2: A Swimmer Whose Body Consists of 3 Parallelepipeds: Conversion to One Plane Motion If the centers of mass of the swimmer's body parts do not lie in the plane of desirable motion, then the swimmer can apply a rotational 2D motion as in Case 1 (see Figs. 11.7, 11.8) to move all the aforementioned centers of mass on the same line and then act as in Case 1 in any desirable plane of motion that goes through the aforementioned line, and so on.

Case 3: A Swimmer Whose Body Consists of 3 Parallelepipeds: The General Case Consider the swimmer in Figs. 11.17, 11.18.

If $z_i(t)$, $i = 1, 2, 3$, do not lie in the plane of desirable motion, then, in order to *change the immediate plane of $z_i(t)$'s toward the desirable one*, the swimmer can use its geometric controls to orient its body parts as follows:

- The parallelepiped with the center at $z_3(t)$ can have its longest dimensions parallel to the desirable motion to the left to ensure its motion in this direction, due to the rotational forces. The side of dimension q does not have to be perpendicular to the plane of the given rotational forces. The other two parallelepipeds can have the longest sides perpendicular to the respective rotational forces, but not so for the sides of dimension q.
- The sides of dimension q of all body parts can also be oriented, relative to the motion to the left along their longest sides, in such a way that, according to Theorem 10.3, they will also move laterally.
- The above actions can momentarily change the plane of the forward motion of our swimmer to the left, see Figs. 11.17, 11.18. (Again, it is difficult to make figures of 3D bodies with all forces in the 2D format of this monograph.)

Case 4: A Particular Example of a Swimmer Whose Body Consists of 6 Parallelepipeds In Fig. 11.19, we show a 3D swimmer with 6 body parts that are small identical parallelepipeds with dimensions $\ell << q \le p$ as in Theorem 10.3.

- The first 3 body parts are oriented as in Case 1 in the above with their centers of mass in one plane, and

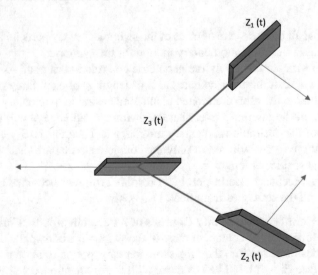

Fig. 11.17 3D breaststroke locomotion: forces acting outside of an incompressible medium

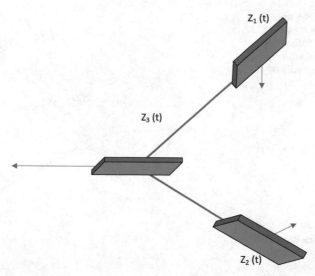

Fig. 11.18 3D breaststroke locomotion: transformed forces from Fig. 11.17 acting inside of an incompressible medium, $z_1(t)$ and $z_2(t)$ also rotate in the plane perpendicular to the direction of the motion to the left, changing the plane of momentary forward motion to the left

Fig. 11.19 A swimmer with 6 body parts as in Sect. 11.5

- the remaining 3 body parts are similarly oriented but in a different plane.
- This swimmer can, potentially, momentarily move the first 3 parts in one plane and the other 3 in a different one, which can result in a momentary motion of its center of mass in any desirable direction in 3D.

Of course, with such a swimmer with 6 body parts, the available set of strategies is more diverse, but also more complicated in implementation, relative to the case of a swimmer with 3 body parts.

11.6   Breaststroke Locomotion of a Swimmer Consisting of 3 Balls in 3D

- In Fig. 11.20, we show a possible set of internal rotational and elastic forces and geometric controls (namely, the same set that would ensure a locomotion of a swimmer consisting of 3 parallelepipeds), engaged by our swimmer outside an incompressible fluid.

Fig. 11.20 Breaststroke
pattern for a swimmer
consisting of 3 balls: Forces
acting outside of an
incompressible medium

Fig. 11.21 Breaststroke
pattern for a swimmer
consisting of 3 balls:
Transformed forces from
Fig. 11.20 acting inside of an
incompressible medium. The
sum of all forces remains to
be zero: no locomotion

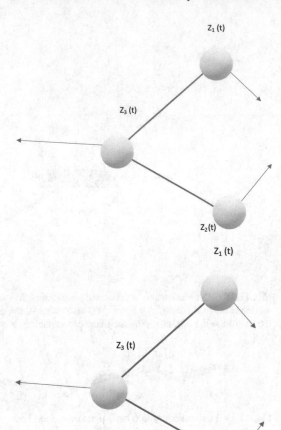

- In Fig. 11.21, we show the actual forces that will act on each ball inside an incompressible fluid: all forces retain the same directions but are reduced in their magnitude by the same factor, as stated in Theorem 10.5.
- The swimmer's center of mass does not move: the sum of all forces acting on the center of mass remains to be zero.

Remark 11.3 (Open Questions and Research Perspectives) The above illustrating examples provide a starting point and guidelines for a rigorous investigation of controllability of various specific bio-mimetic swimmers and living organisms along Definitions 11.1 and 11.2 and Remarks 9.4, 10.7, and 11.2. Further research can also target models which include some known specific external forces acting upon their bodies.

References

1. F. Alouges, A. DeSimone, A. Lefebvre, Optimal strokes for low Reynolds number swimmers: An example. J. Nonlinear Sci. **18**, 277–302 (2008)
2. L.E. Becker, S.A. Koehler, H.A. Stone, On self-propulsion of micro-machines at low Reynolds number: Purcell's three-link swimmer. J. Fluid Mech. **490**, 15–35 (2003)
3. P. Cannarsa, A.Y. Khapalov, Micromotions and controllability of a swimming model in an incompressible fluid governed by 2-D or 3-D Navier–Stokes equations (with P. Cannarsa). J. Math. Anal. Appl. (JMAA) **465**, 100–124 (2018). https://doi.org/10.1016/j.jmaa.2018.04.066
4. S. Childress, *Mechanics of Swimming and Flying* (Cambridge University Press, 1981)
5. R. Cortez, The method of regilarized stokeslets. SIAM. J. Sci. Comput. **23**, 1204–1225 (2001)
6. Gi. Dal Maso, A. DeSimone, M. Morandotti, An existence and uniqueness result for the motion of self-propelled micro-swimmers. SIAM J. Math. Anal. **43**, 1345–1368 (2011)
7. T. Fakuda et al., Steering mechanism and swimming experiment of micro mobile robot in water, in *Proc. Micro Electro Mechanical Systems (MEMS'95)*, pp. 300–305 (1995)
8. L.J. Fauci, Computational modeling of the swimming of biflagellated algal cells. Contemporary Mathematics **141**, 91–102 (1993)
9. L.J. Fauci, C.S. Peskin, A computational model of aquatic animal locomotion. J. Comput. Phys. **77**, 85–108 (1988)
10. G.P. Galdi, On the steady self-propelled motion of a body in a viscous incompressible fluid. Arch. Ration. Mech. Anal. **148**(1), 53–88 (1999)
11. G.P. Galdi, On the motion of a rigid body in a viscous liquid: A mathematical analysis with applications, in *Handbook of Mathematical Fluid Mechanics*, ed. by S. Friedlander, D. Serre (Elsevier Science, 2002), pp. 653–791
12. J. Gray, Study in animal locomotion IV - the propulsive power of the dolphin. J. Exp. Biol. **10**, 192–199 (1032)
13. J. Gray, G.J. Hancock, The propulsion of sea-urchin spermatozoa. J. Exp. Biol. **32**, 802 (1955)
14. S. Guo et al., Afin type of micro-robot in pipe, in *Proc. of the 2002 Int. Symp. on Micromechatronics and Human Science (MHS 2002)*, pp. 93–98 (2002)
15. M.E. Gurtin, *Introduction to Continuum Mechanics* (Academic Press, 1981)
16. V. Happel, H. Brenner, *Low Reynolds Number Hydrodynamics* (Prentice Hall, 1965)
17. S. Hirose, *Biologically Inspired Robots: Snake-Like Locomotors and Manipulators* (Oxford University Press, Oxford, 1993)
18. E. Kanso, J.E. Marsden, C.W. Rowley, J. Melli-Huber locomotion of articulated bodies in a perfect fluid. J. Nonlinear Sci. **15**(4), 255–289 (2005)

19. A.Y. Khapalov, The well-posedness of a model of an apparatus swimming in the 2-D Stokes fluid, Techn. Rep. 2005-5, Washington State University, Department of Mathematics, Tech. Rep. Ser., http://www.math.wsu.edu/TRS/2005-5.pdf
20. A.Y. Khapalov, Local controllability for a swimming model. SIAM J. Control. Optim. **46**, 655–682 (2007)
21. A.Y. Khapalov, Global controllability for a swimming model, submitted, also available as Techn. Rep. 2007–11, Washington State University, Department of Mathematics, Tech. Rep. Ser., http://www.math.wsu.edu/TRS/2007-11.pdf
22. A.Y. Khapalov, Geometric aspects of force controllability for a swimming model. Appl. Math. Opt. **57**, 98–124 (2008)
23. A.Y. Khapalov, Micro motions of a 2-D swimming model governed by multiplicative controls. Nonlinear Anal. Theory Methods Appl. Special Issue WCNA 2008 **71**, 1970–1979 (2009)
24. A.Y. Khapalov, Swimming models and controllability, in *Proc. Int. Conf. (Fes, Morocco) Sys. Theory: Modeling, Analysis and Control*, pp. 241–248 (2009)
25. A.Y. Khapalov, *Controllability of Partial Differential Equations Governed by Multiplicative Controls*. Lecture Notes in Mathematics Series, Vol. 1995 (Springer, Berlin, Heidelberg, 2010), 284p.
26. A.Y. Khapalov, Wellposedness of a swimming model in the 3-D incompressible fluid governed by the nonstationary Stokes equation. Int. J. Appl. Math. Comput. Sci. **23**(2), 277–290 (2013). https://doi.org/10.2478/amcs-2013-0021
27. A.Y. Khapalov, Micro motions of a swimmer in the 3-D incompressible fluid governed by the nonstationary Stokes equation. SIAM J. Math. Anal. (SIMA) **45**(6), 3360–3381 (2013). http://epubs.siam.org/toc/sjmaah/45/6. https://doi.org/10.1137/120876460
28. A.Y. Khapalov, Addendum to "Wellposedness of a swimming model in the 3-D incompressible fluid governed by the nonstationary Stokes equation". Int. J. Appl. Math. Comput. Sci. (AMCS) **23**(4), 905–906 (2013). https://doi.org/10.2478/amcs-2013-0067
29. A.Y. Khapalov, Wellposedness results for swimming models in the 2-D and 3-D incompressible fluids described by the nonstationary Stokes equation, Washington State University, Department of Mathematics, Techn. Rep. Ser., 2014-1, http://www.math.wsu.edu/TRS/2014-1.pdf
30. A.Y. Khapalov, *Mobile Point Sensors and Actuators in the Controllability Theory of Partial Differential Equations* (Springer International Publishing, 2017), 233p.
31. A.Y. Khapalov, Sh. Eubanks, The well-posedness of a 2-D swimming model governed in the nonstationary Stokes fluid by multiplicative controls. Applicable Analysis **88**, 1763–1783 (2009)
32. A.Y. Khapalov, G. Trinh, Geometric aspects of transformations of forces acting upon a swimmer in a 3-D incompressible fluid. Disc. Cont. Dyn. Sys. A **33**(4), 1513–1544 (2012)
33. A.Y. Khapalov, P. Cannarsa, F. Priuly, G. Floridia, Well-posedness of 2-D and 3-D swimming models in incompressible fluids governed by Navier–Stokes equations. J. Math. Anal. Appl. **429**, 1059–1085 (2015). https://doi.org/10.1016/j.jmaa.2015.04.044
34. J. Koiller, F. Ehlers, R. Montgomery, Problems and progress in microswimming. J. Nonlinear Sci. **6**, 507–541 (1996)
35. A.N. Kolmogorov, S.V. Fomin, *Elements of the Theory of Functions and Functional Analysis* (Imprint Rochester, NY, Graylock Press, 1957)
36. O.H. Ladyzhenskaya, *The Mathematical Theory of Viscous Incompressible Flow* (Cordon and Breach, New York, 1969)
37. O.H. Ladyzhenskaya, *Mathematical Questions of the Dynamics of the Viscous Incompressible Fluid* (Nauka, Moscow, 1970)
38. O.H. Ladyzhenskaya, V.A. Solonikov, N.N. Ural'ceva, *Linear and Quasi-linear Equations of Parabolic Type* (AMS, Providence, Rhode Island, 1968)
39. L.G. Leal, *The Slow Motion of Slender Rod-Like Particles in a Second- Order Fluid, J. Fluid Mech.*, vol. 69, pp. 305–337 (1975).
40. M.J. Lighthill, *Mathematics of Biofluiddynamics* (Society for Industrial and Applied Mathematics, Philadelphia, 1975)

41. S. Martinez, J. Cortés, Geometric control of robotic locomotion systems, in *Proc. X Fall Workshop on Geometry and Physics, Madrid, 2001, Publ. de la RSME*, vol. 4, pp. 183–198 (2001)
42. R. Mason, J.W. Burdick, Experiments in carangiform robotic fish locomotion, in *Proc. IEEE Int. Conf. Robotics and Automation*, pp. 428–435 (2000)
43. K.A. McIsaac, J.P. Ostrowski, Motion planning for dynamic eel-like robots, in *Proc. IEEE Int. Conf. Robotics and Automation*, San Francisco, pp. 1695–1700 (2000)
44. V.P. Mikhailov, *Partial Differential Equations* (Mir, Moscow, 1978)
45. K.A. Morgansen, V. Duindam, R.J. Mason, J.W. Burdick, R.M. Murray, Nonlinear control methods for planar carangiform robot fish locomotion, in *Proc. IEEE Int. Conf. Robotics and Automation*, pp. 427–434 (2001)
46. C.S. Peskin, Numerical analysis of blood flow in the heart. J. Comput. Phys. **25**, 220–252 (1977)
47. C.S. Peskin, D.M. McQueen, A general method for the computer simulation of biological systems interacting with fluids, in *SEB Symposium on biological fluid dynamics*, Leeds, England, July 5–8 (1994)
48. J. San Martin, J.-F. Scheid, T. Takashi, M. Tucsnak, An initial and boundary value problem modeling of fish-like swimming. Arch. Ration. Mech. Anal. **188**, 429–455 (2008)
49. A. Shapere, F. Wilczeck, Geometry of self-propulsion at low Reynolds number. J. Fluid Mech. **198**, 557585 (1989)
50. M. Sigalotti, J.-C. Vivalda, Controllability properties of a class of systems modeling swimming microscopic organisms. ESAIM: COCV. https://doi.org/10.1051/cocv/2009034. Published online August 11, 2009
51. W.I. Smirnow, *Lehrgang der h"heren Mathematik. Teil IV/1* (Translated from the Russian. Hochschulbucher fur Mathematik, 5a. VEB (Deutscher Verlag der Wissenschaften, Berlin, 1988), 300 pp.
52. H. Sohr, *The Navier–Stokes Equations*. An Elementary Functional Analytic Approach (Birkhäuser, 2001)
53. K.R. Symon, *Mechanics*. Addison-Wesley Series in Physics (1971)
54. G.I. Taylor, Analysis of the swimming of microscopic organisms. Proc. R. Soc. Lond. A **209**, 447–461 (1951)
55. G.I. Taylor, Analysis of the swimming of long and narrow animals. Proc. R. Soc. Lond. A **214**, 158–183 (1952)
56. R. Temam, *Navier-Stokes Equations* (North-Holland, 1984)
57. M.S. Trintafyllou, G.S. Trintafyllou, D.K.P. Yue, Hydrodynamics of fishlike swimming. Ann. Rev. Fluid Mech. **32**, 33–53 (2000)
58. T.Y. Wu, Hydrodynamics of swimming fish and cetaceans. Adv. Appl. Math., **11**, 1–63 (1971)

Printed in the United States
by Baker & Taylor Publisher Services